Initial Professional Development for Civil Engineers

Initial Professional Development for Civil Engineers

H. Macdonald Steels

Published by ICE Publishing, One Great George Street,
Westminster, London SW1P 3AA.

Full details of ICE Publishing sales representatives and distributors can be found at:
www.icevirtuallibrary.com/info/printbooksales

First published 2011
Reprinted 2014

Other titles by ICE Publishing:
Dynamic Mentoring for Civil Engineers. H. M. Steels.
ISBN 0-7277-3003-7
Effective Training for Civil Engineers. H. M. Steels.
ISBN 0-7277-2709-5
Successful Professional Reviews for Civil Engineers, 3rd edition.
H. M. Steels. ISBN 978-0-7277-4152-3

www.icevirtuallibrary.com

A catalogue record for this book is available from the British Library

ISBN 978-0-7277-4147-9

ICE Publishing is a division of Thomas Telford Ltd, a wholly-owned subsidiary of the Institution of Civil Engineers (ICE).

Typeset by Academic + Technical, Bristol
Index created by Indexing Specialists (UK) Ltd, Hove, East Sussex
Printed and bound in Great Britain by CPI Antony Rowe
Limited, Chippenham and Eastbourne

Contents

Foreword

The framework for Initial Professional Development is established by the ICE3000 series of documents *Routes to Membership* (currently amendment A), each of which can be accessed direct at www.ice.org.uk/ICE3000A, etc.

The Value of Professional Qualification	ICE3000A
Routes to Membership (CEng and IEng)	ICE3001A
Route to Technician Membership	ICE3002A
Route to Associate Membership	ICE3003A
Individual Routes to Membership	ICE3004A
Development Objectives (all routes)	ICE3005A
Continuing Professional Development	ICE3006A

Additional ICE advice and guidance is available from the Membership Guidance Notes (MGN), which may be accessed in a similar manner at www.ice.org.uk/mgn followed by the number of the MGN. The following MGNs are relevant to most chapters:

Written Assignment Topic Areas	MGN 21 www.ice.org.uk/mgn21
Written Test Topic Areas	MGN 22 www.ice.org.uk/mgn22

Acknowledgements

The basis of this book on professional development for graduate civil engineers is learning through experience; not just their own, but mainly through the experience of other people.

The content is proof of that concept. It could not have been written without the experience gained from so many people over so many years. Many of them probably did not realise the enormous influence they were having on my development; they merely offered informal advice, encouragement and opinions when they felt it necessary or whenever I asked.

This willingness to share experience through discussion, one of the most admirable qualities of our profession, is one which new entrants will find of inestimable value if they tap into it. No one is going to formally 'train' them, but nearly all engineers will readily respond with guidance, advice and, above all, experience, whenever they are approached.

I cannot list all the people whose experience has helped me during my career, but there are some whom I must mention. Lindsey Ellis, who has shared her extensive knowledge and understanding of the training process freely and with great good humour over many years. Developing and refining seminars with Patrick Waterhouse for Bowdon Consulting provided much of the content of this book. My elder son Duncan's incisive criticism fundamentally shaped the format.

I owe them, and myriad others, a huge debt of gratitude. Through their experience, I have learnt a very great deal. Any errors in this book cannot be attributed to them; they result solely from my misunderstandings. All opinions are also mine. If you disagree with them, it just shows that you are thinking about the broader issues and developing your own views – you are becoming a professional civil engineer.

'I keep six honest serving-men
(They taught me all I knew);
Their names are What and Why and When
And How and Where and Who.'

Rudyard Kipling (1902), the first four lines of *The Elephant's Child*

Initial Professional Development for Civil Engineers
ISBN 978-0-7277-4147-9

doi: 10.1680/ipdce.41479.001

Chapter 1
Introduction

Initial Professional Development

This book provides a 'foundation' of information on skills, knowledge and understanding on which the development of all civil engineers, and their preparation for Professional Reviews, is based. It will help candidates and their mentors to make sense of the process, giving them insight into how to review their experience and the learning they must take from it. More mature candidates, who have come to a later decision to progress to Membership at either IEng or CEng, can readily catch up by using the content. The book offers all Review candidates the vision to select the key, important elements of experience to demonstrate their understanding, skills, knowledge and insight.

Initial Professional Development (IPD) is defined by the Institution of Civil Engineers (ICE) as 'the acquisition and development of the special skills and professional approach needed to practise as a Civil Engineer'. The ICE publication *ICE3001 – Routes to Membership* goes on to state: 'Acquiring learning and experience at work will develop your ability to hold positions of responsibility and make independent judgements'. So, IPD is a period of structured training which usually follows, or is contiguous with, academic education and leads to qualification through a Professional Review.

The brief descriptions given in the ICE3000 series of documents are the culmination of many years' experience of trying to define the nebulous skills and abilities of professional civil engineers. Rather like 'management', there is no precise definition and, in any case, the fundamental concept changes with time, context and circumstances. The current descriptors are, arguably, the best attempt so far, but are unlikely to be the last.

Appendix A of ICE3001, first published in February 2006, concisely describes nine attributes which the Institution requires potential Members to demonstrate at Professional Review. The central column expands the attributes for Membership, the right-hand column stretches them further for Chartered Membership. These attributes form the skeleton of this book; each is fleshed out so that potential professional engineers better understand their personal target. The majority of the text is applicable to all potential Members, whether recognised as Incorporated or Chartered by the Engineering Council. The extrapolations for a Chartered Engineer are added as appropriate in each chapter.

Attributes 1 and 2 for Associate Membership (ICE3003) are different, with their requirement for scientific and technical rather than engineering skills but, nevertheless, the advice given is equally applicable.

How this book is organised

> The advice and guidance from the Institution, as well as statutes, standards, laws and regulations with which we have to comply, are being continually updated. Therefore, all documents are referred to generically, without dates or amendments. So ICE3001(A), current at the time of writing, is still referred to in the text as ICE3001. Readers must ensure that they have the latest edition of any document.

Chapter 2 discusses the role and responsibilities which are the target – to become a professional civil engineer. The Professional Reviews are holistic, not examining each attribute in detail, but reviewing your overall capability as a civil engineer.

Chapters 3 and 4 discuss the transition from educational learning to workplace learning. For some this step seems relatively easy, but for many others it takes some time to realise the fundamental reasons for their difficulties; valuable time is therefore 'lost' and their initial professional development can take longer than it might otherwise have done.

ICE Review statistics show that attributes 1 and 2 cause difficulties for too many candidates, so Chapter 5 explores possible reasons for this and offers guidance on the proper interpretation of these two technical attributes.

Chapters 6 and 7 cover the personal attributes of leadership, management, independent judgement and responsibility.

The Membership Guidance Notes (MGNs), particularly MGNs 21 and 22, indicate that the Commercial attribute (5) covers a very wide spectrum of knowledge and understanding, essentially encompassed by the words 'statutory and commercial frameworks', which consequently spill over into two further chapters – 9 and 10.

Finally, the remaining attributes are covered in the order in which they appear in Appendix A, concluding with what is, in effect, a summary of the book in Professional Commitment (attribute 9).

Each chapter opens by drawing together all published information from the ICE about that particular attribute, thus presenting a comprehensive framework of ICE information within which trainees can develop their own understanding. This includes the attribute descriptors, together with excerpts from relevant Development Objectives and MGNs, as appropriate.

There is no set sequence of progression. Each individual's route will be determined by

- their own particular strengths and weaknesses
- their personal preferences
- the experience available
- the stage in procurement in which they become involved.

As a consequence, each chapter is complete in itself, referenced to the others as appropriate. This does lead to a certain amount of repetition, but this has been kept to a minimum.

The Development Objectives

The Institution provides a framework of Development Objectives (ICE3005), designed to assist in the logical and progressive development of the requisite attributes. There is, however, no direct correlation between the ICE Development Objectives and the personal attributes (ICE3001) to be developed, although there is considerable overlap and interrelationship. This only goes to confirm the complexity of civil engineering and the need for civil engineers to deploy many talents simultaneously but in diverse combinations. However, certain Development Objectives can be loosely related to the attributes, as suggested at the beginning of each attribute chapter.

The Objectives are only a means to an end, not an end in themselves. So do not fall into the trap of wanting to have 'Objectives signed off'; that wish would suggest a 'box-ticking' attitude, which is certainly not what is required. The outcome is intended to be the development of the 'special skills and professional approach' needed to practise.

It is for this reason that the Objectives form no part of the Professional Review; by then they have been assimilated. It is evidence of adequate capability in the nine attributes (the product) which you must demonstrate. Formal completion of IPD (the process) must be signified prior to your Review, either by your Supervising Civil Engineer if under Agreement or by the Institution through a Career Appraisal, with exceptions only for the Technical Report Route and the Senior Route.

Experience required for IPD

Each attribute is a complex mix of knowledge, skills and understanding, which can only be developed by continuous practice, reading and thinking. But there is no prescribed experience. Each trainee, with help from their employer and colleagues, must decide for themselves just what is necessary and how sufficient experience might be gained within their particular working environment. This book will help you in making those decisions.

As an example, there has been, for many years, widespread confusion as to whether every civil engineer *must* have a mix of at least 12 months' design and 12 months' site experience. That has not been a *requirement* for years. Indeed, the Institution felt it necessary to write MGN 42, setting out very clearly what criteria related to this aspect will be used to evaluate a candidate for membership.

After all, every UK professional civil engineer is required by law to know and fully understand the implications of their decisions and actions for those following them in the procurement process. We must all envisage how our successors can carry out their responsibilities and, wherever possible, make their involvement as safe as we can by minimising hazards and limiting risks, identifying any significant residual risks for those who succeed us. This ability will certainly be thoroughly tested at Professional Review.

So you must develop an adequate understanding of the whole procurement process, through site investigation, feasibility, resource allocation, design, construction, use, maintenance and eventual reconstruction or demolition. You are strongly advised not only to read MGN 42, but also to understand the philosophy behind it, because that fundamental attitude is applicable to all other aspects of your training and development. It is not what you do or for how long you do it, but what you learn from your experiences, how you develop the attributes, that are important. For some engineers, an extended period in construction (or implementation) might be necessary; at the other extreme, for other candidates this practical part of the procurement process is almost second nature.

Initial Professional Development for Civil Engineers
ISBN 978-0-7277-4147-9

doi: 10.1680/ipdce.41479.005

Chapter 2
What is a Professional Civil Engineer?

Risk management

Some years ago now, I was being interrogated by a lawyer, who asked, 'Is this building safe?'

'It is perfectly safe', I replied.
'So it will not fall down?'
'Yes it could!'
'But you've just told me it is perfectly safe.'
'That is correct', I replied.
'So why do you say it could fall down?'
'Because I have little idea how it would perform if the presumed loading conditions are exceeded.'

We went on to explore how the design parameters appropriate for that particular high-rise building had been decided – in terms of wind, snow, rain, people, equipment and storage, road traffic impact, ground bearing capacity, earthquake and, since it was under the flight path to an airfield, aircraft impact.

This brief interchange highlights that the civil engineer's fundamental business is risk management. We must visualise all possible hazards (not merely the technical ones) and then decide which risks are so slim that they can be ignored, which must be designed against and which could be avoided. One of the chief designers of the World Trade Center in New York, Leslie Robertson, said he had considered the possible impact of a lost Boeing 707 jet airliner trying to land in fog at relatively low speed at the international airports (amongst several aircraft incidents in New York, a B25 bomber *had* hit the 79th floor of the Empire State Building in July 1945). But, significantly, because the threat was seen as a landing aircraft, there was no consideration of the possible effects of impact at higher speeds with full fuel tanks. Subsequent Inquiries found that this was a primary cause of the fires and the collapses of the twin towers on September 11 2001, but all agreed that it was a totally unforeseeable hazard, which could not have been designed against.

It is not only possible threats to structural safety that have to be considered. Civil engineers must also anticipate all possible effects which their proposals could have on economics, flora and fauna, the Earth's resources, global stability and climate change, the users, maintainers and demolishers of any facility and everybody and everything affected in any way by the proposal. A tall order indeed!

We have a whole host of documents setting out established best practice – codes and standards giving guidance on what has been found to work previously. These form the basis for most of our decisions. But they are neither absolute nor can they foresee potential changes or events. Compliance with established practice is not enough; civil engineers and others involved in the built environment must consider whether existing practice is still adequate, or whether conditions have strayed beyond those experienced previously.

Institution descriptions

At the start of the new millennium, the Institution produced a new 'definition' of civil engineering, attempting to bring up to date the original 1818 description, which distinguished civil from military engineering. Part of the new description stated that:

> Its essential feature, as distinct from science and arts, is the exercise of imagination to fashion the products, processes and people needed to create a sustainable built environment. It requires a broad understanding of scientific principles, a knowledge of materials, and the art of analysis and synthesis. It also requires research, team working, leadership and business skills. (*New Civil Engineer*, 2007)

The 'exercise of imagination' is perhaps not something that is readily apparent during your education, certainly not as the 'essential feature' of civil engineering. Yet, it is vital that civil engineers can visualise the problems, imagine what solutions might be possible and, just as important, what resources might be required. They must also imagine the effects that the solution and the obtaining and transporting of the resources would have on the entire environment. What would be the effect of using these scarce resources (only timber is a renewable engineering material) on future generations trying to meet their needs (which is what sustainability is all about)? Every one of these judgements should be made for every project, from patch repair to multi-million pound developments. As an example, a graduate group was unable to tell me how their road repairs affected the West Indies, even though we were standing beside a drum marked 'Trinidad Lake Asphalt'! So imagination *is* undoubtedly the essential feature!

The definition continues by making a distinction between civil engineering and both science and arts. It is neither an exact, defined science nor a 'skill applied to music, painting, poetry or sculpture' (*Collins Dictionary* definition of 'arts'). Yet the definition then goes on to say that civil engineering requires 'a broad understanding of scientific principles' and 'the art of analysis and synthesis'! What do these apparent contradictions mean?

Analysis and synthesis

Analysis is the separation or breaking down of any problem into its constituent parts or elements. It is the examination of these component parts, both separately and in relationship to the whole. Much of a civil engineer's mathematical analysis is based on previously determined best practice (i.e. that which has been proven to give realistic results). So, it is akin to science, but does not have the same exactitude. Many of the parameters are not capable of precise definition, but require us to make sensible judgements – an art based on a 'broad understanding of scientific principles'. So we can analyse a beam or a column using scientific principles, but based on judgements about the nature and likely performance of concrete and steel and their interaction. This is why we use factors of safety – to allow for inexactitude.

Synthesis is the combining of separate simple elements into a complex whole. The opposite of analysis, it requires us to look at how each piece of analytical thought affects and is affected by the other pieces of analysis. So, in my simple example, the beam will interact with the column and the column will have an effect on the stiffness of the beam.

A much more complex example is the environment, each item of which can be examined in considerable detail in isolation, using scientific principles to replicate its behaviour. But every part of the environment affects every other part in a complex matrix of interaction, which is extremely difficult to comprehend, let alone quantify.

Appendix A of ICE3000 is, arguably, an example of analysis, offering a detailed breakdown of what ICE means by 'the special skills and professional approach you will need to practise'. Candidates still ask me what the relative weighting is of each of the attributes (i.e. which are most important and which less so). There is no weighting. *All* are important constituents of a professional civil engineer. The Reviewers are asked to make a holistic judgement on the synthesis of these attributes.

The holistic engineer

Each candidate is reviewed as a complete person and a judgement is made on whether or not they are capable of carrying the responsibilities of a professional civil engineer – 'Would this person, when placed in a position of responsibility, make the correct decisions?' Only after answering that question do they break their judgement down into the attributes, to determine the sources of any doubts and uncertainties. By doing this, they can offer some guidance as to how an unsuccessful candidate could rectify their perceived shortcomings for a future attempt. The process is not dissimilar to the UK driving test, where the first question the examiner answers is 'Did I feel safe?' Only then do they complete their forms to explain why.

So, it is possible (and it has happened, although I would not recommend any candidate to rely on it!) that a Review candidate can be deemed inadequate in a specific attribute but still be successful at Review. The Reviewers must be confident that the candidate would properly address that inadequacy without prompting, should the need ever arise. The only current exception is Health, Safety and Welfare, which, understandably given the nature of the risks we assess, must be at least adequate.

Civil engineers are in demand in many spheres of employment, well beyond the built environment. It is worthwhile examining why. *Prospects*, the official UK graduate careers website, states that 'the skills and qualities nurtured by studying civil engineering suit many other professional settings equally well'. It goes on to list those skills and qualities

- creativity and an innovative approach to solving problems
- the ability to analyse and interpret diverse, complex data
- critical thinking and the ability to evaluate designs, plans and projects
- effective assessment and management of risk, resources and time
- highly developed numeracy and computer literacy
- interpersonal sensitivity, persuasiveness and the ability to work as part of a team
- clear written and oral communication skills
- awareness of ethical issues and the wider impact of your work.

This is a very complimentary description of what we are, but it also hints at a great responsibility – to judge what is appropriate, based on a broad understanding of engineering, society and the planet.

Understanding

This word 'understanding' appears frequently throughout the Institution's documentation ('a broad understanding of scientific principles', 'a sound understanding of core engineering principles', 'high level of commercial and contractual understanding', 'sound understanding of the construction process') and throughout this book. What does it actually mean?

During the construction of the Konkan railway from Mumbai to Mangalore on the west coast of India, a poem was found pinned to a site hut (author unknown). It read (translated):

> I take the vision which comes from dreams
> And apply the magic of science and mathematics;
> Adding the heritage of my profession
> And my knowledge of nature's materials,
> To create a design.
>
> I organise the efforts and skills of my fellow workers,
> Employing the capital of the thrifty

And the products of many industries.
Together we work towards our goal,
Undaunted by hazards.

And, when we have completed our task,
All can see that the dreams have materialised
For the comfort and welfare of all.
I am an engineer; I serve mankind.
I make dreams come true.

While painting a rather romantic image, it nonetheless does portray a profession that is not solely the application of rules and formulae, but one which is reliant on judgement and a broad understanding of so many facets of the natural world and the aspirations of mankind.

Civil engineering is a logical process of making judgements, which are based on some fundamental laws, rules and concepts. It is, by no means, an exact science, so the manipulation of actual or contrived facts, using equations and formulae, cannot alone provide adequate answers. Mechanistic analysis, often based on published previous best practice in codes and standards, is a fundamental tool in the design process. However, it is only a small (albeit critical) part of the whole procurement process. It is where we satisfy ourselves that the solution we are proposing will work satisfactorily under predetermined conditions.

Judgement

The majority of the procurement process is judgement, and this requires a deep and comprehensive understanding of the many factors that influence those judgements. A High Court judge first defined a civil engineer's job for me in these terms: 'at this time, with these resources, in these circumstances, for the foreseeable future, this is the best I can do'. This is a very strong decision and it can be defended, even in a court of law. But it is not absolute. One or more of those parameters is bound to change.

So, engineering is a balancing of many conflicting parameters to achieve a workable solution to what is usually a complex problem. This balance can be simplistically likened to an infinitely flexible three-dimensional membrane, being pulled in all directions by a whole range of conflicting considerations (Figure 1).

Each of these considerations is, itself, another membrane of conflicting parameters.

- 'Function' means that the chosen solution must work (i.e. it must do what it has been decided is required – a decision which is, in itself, another membrane). This is where analytical calculations are needed. But this consideration is now of far

Figure 1

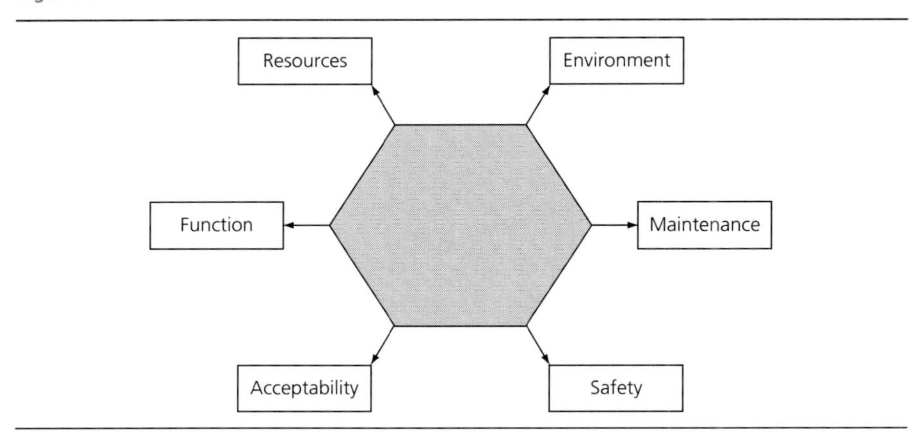

greater importance than in the days of 'predict and provide'. Now, we must curb excessive demand, and focus on satisfying peoples' needs rather than their wants or desires (see Chapter 12). In the UK's developed society, we are moving into an era of 'manage and maintain'; trying to make the existing infrastructure work more efficiently, with improvements only where necessary – another membrane.

- 'Resources' is a balancing membrane between what can be referred to as the 'five Ms' – manpower, machines, methods, materials, all dependent on money. Each of these is affected by availability, cost and suitability (e.g. money has to be borrowed or diverted from other purposes, materials are predominantly finite and must be conserved and methods are highly dependent on energy, largely from diminishing fossil fuels).
- 'Environment' is a hugely complex membrane in its own right. Everything we do affects the environment and we must, more than ever before, balance the improvements we are making against the damage we are causing. The concept of sustainability (Chapter 12) requires us to question and limit the continuing, prolific use of non-renewable resources, while a greater awareness of the vulnerability of flora and fauna is forcing a far greater emphasis on their survival and well-being.
- 'Safety' is a balance between the hazards which we are prepared to design to resist and the resources we might be able to deploy to eliminate or reduce the risks. With weather patterns apparently becoming more extreme, and with small militant groups able to commit ever-greater acts of destruction, both in the UK and elsewhere this factor necessitates a balance between risk aversion and the cost to life, limb, environment and bank balance of a catastrophic event.
- 'Maintenance' is a balance between the original specification and the repair needed to retain the effective use of the asset. A 'design life' of 100 years does not presume that the asset will last without attention for 100 years, but that the risk of

maintenance being required will not exceed limits defined at the design stage. For example, a bridge may be expected to have a life of 100 years, but no manufacturer is going to guarantee the bearings for that length of time. So provision, financial and practical, must be made for replacing the bearings during the life of the bridge.

- 'Acceptability' is dependent on satisfying the majority of the population (whether directly affected or not) that the balance which is being promoted is in their best interests. In today's world, it is extremely difficult to get the unbiased facts of the project into the public domain, to counter the strident and 'newsworthy' criticism of pressure groups. Significant resources must be put into careful, considered and comprehensive explanations of how and why the proffered balance has been determined, if any infrastructure proposal is to be acceptable.

Think about each of the projects with which you are involved. How were each of the balances achieved for them? Do you think they were the best solutions to that particular set of circumstances? This curiosity requires you to lift yourself away from the job and consider the wider implications – the 'helicopter factor'. Hover ever-higher above your immediate responsibilities to enquire about

- your colleagues (their contribution)
- your group or section (other projects in the group)
- the organisation (strategic direction, resource allocation)
- the client (political/commercial 'will', purposes and direction)
- central and local government (laws, policies, often financial implications)
- Europe (policies, grants)
- the World (environment, trade agreements, international targets).

The one overriding characteristic of all these considerations is that they are not constant. They are dependent on

- shifting public opinion
- fickle political will
- emerging national wealth
- changing global circumstances, for example:
  - for many years it may have been sufficient to design sewers or flood plains for a certain hypothetical storm but, with more extreme weather patterns becoming apparent, is this still adequate?
  - should coastal erosion be allowed or should the present shape of the land be defended?
  - expenditure which may be economic for a wealthy nation may not be realistic for an emerging nation
  - the priorities for expenditure are dependent on the aspirations of the people in control of the expenditure

- the sophistication of a given solution must match the current level of development of the people in that locality
- the value to the public of a human life, energy, transport, a species or a rainforest will vary depending on the current attitudes and aspirations of the area's inhabitants.

To achieve these balances, civil engineers must be able to do a lot more than merely comply with established best practice. They must

- be aware of and relate to constantly changing circumstances
- be able to reduce each problem to its simplest components
- identify the principles that make each component work
- make logical connections between the many differing components and circumstances
- assess the effects of each component on the others
- extrapolate from what they already understand to something new which is encountered.

It is the amalgam of such abilities which comprises 'understanding'. Understanding allows knowledge to be put to best use. Knowledge is the awareness of discrete pieces of information; understanding is what enables those pieces to be synthesised. Essentially, it is the ability to conceptualise, sometimes referred to as critical insight. For example

- you understand a mathematical concept if you are able to use it to solve problems which are unlike problems you have met before
- you understand a foreign language if you can not only translate the words into your native tongue, but also reproduce the attitudes, information and nuances of the message
- you understand the environment if you can successfully imagine or visualise what effects your proposals could have on it.

You understand an aspect of engineering if you are able to do *all* of the following

- easily make logical connections between different facts and concepts
- identify the principles of engineering which make everything work
- recognise the connections between the engineering that you understand and something new which you encounter
- explain the concepts and facts in simple terms to people who do not have a knowledge of the subject.

As part of the decision-making process, analytical techniques are used, based on fundamental laws and principles established over time, to satisfy ourselves that a solution will work under the conditions which we have prescribed. It is these essentially mathematical calculations and processes that form a large part of academic education. Many employers will, quite understandably, require you to apply these techniques at the beginning of your IPD; you will therefore develop a 'sound evidence-based approach to problem solving' and 'maintain and extend a sound theoretical approach to the application of science and technology'.

As your career advances, the work will increasingly require you to 'identify, review and select techniques, procedures and methods' most applicable to your unique problems in order to achieve realistic answers. To do this successfully requires you to understand the technical principles on which they are based, and thus to decide whether those principles equate to the principles underlying the problem. This is the 'understanding of technical (scientific) principles' to be demonstrated at the Professional Reviews. You do not have to be a designer to exhibit this understanding; all civil engineers, at every stage of procurement, must understand the principles behind what is being proposed or implemented if they are to make a sensible contribution.

You will also become involved in defining the problems and assessing the likely effects of alternative solutions, devising plans for their resolution and 'evaluating the effectiveness' of the outcomes. Consequently, you will become increasingly involved in the interface between the built environment (projects created by humans) and the greater world (the natural and social environment). To achieve a balanced solution, judgement is needed, and this must be based on a broad and thorough understanding of all relevant aspects of the built and natural worlds and how they interact.

Initial Professional Development for Civil Engineers
ISBN 978-0-7277-4147-9

doi: 10.1680/ipdce.41479.015

Chapter 3
Learning through experience

Based on my experience of helping many graduates to adapt, there is still a significant gap between academic education and workplace learning. The fundamental cause is the different learning styles which must be used. In the universities, the sheer quantity of material to be transmitted, and the need to build on the educational standards achieved in the schools, necessitate a huge amount of teaching, where students absorb information, make some attempts to utilise it in examples, and then regurgitate it during formal examinations. The learning process is essentially driven by the syllabus and the lecturers, with the students as relatively passive 'receivers' of information, broken down for them into digestible, semester-sized chunks. The available resources are poured into those students who are struggling, so that the university can meet prescribed targets for examination success rates.

Contrast this with the workplace, where there are few 'teachers'. As a graduate, you are expected to learn largely for yourself, to actively manage your own personal development by becoming a 'seeker' of information, either as and when it becomes available or by finding it. What resources there are tend to be concentrated on those who are making good progress, where the value of that investment is gained soonest, rather than on those who are struggling, where any commercial benefits will be longer term.

Mentors often seem surprised that graduates do not respond positively in this environment! Many struggle because the skills that are needed have, of necessity, largely been suppressed by the intense academic learning environment. The perception of learning may be distorted by past experience. Perhaps this is why so many graduates are continually seeking to attend formal courses, trying to return to a familiar learning environment. And yet research consistently shows that lectures and teaching (which comprise the bulk of 'education') are the least successful methods of learning of them all.

Learning through experience

Experience alone teaches us nothing. Newton reputedly formulated the laws of gravity by watching falling apples, which had been falling for thousands of years – there was plenty of experience, but nobody until Newton asked the right questions. Nobody was curious. Nobody asked 'Why?'. Nobody learnt.

Learning is the outcome of curiosity, a desire to know why something happens, why more experienced people make certain decisions, why some people are prepared to take greater risks than others. Here are some methods of learning, with brief explanations

- Education being told
- Research finding out from books, Internet
- Enquiry asking someone
- Discussion talking about it
- Observation watching someone
- Practice having a go.

Generally speaking, graduates come into the workplace with most or all of these skills, but with the more highly developed abilities concentrated at the top of the list. What you must now do, is totally reverse the order, becoming most proficient at discussion, enquiry, observation and – within the strict limits of operational efficiency and risk – practice. Very little of what you will learn from now on will be taught.

You must revert to those learning methods you used, very successfully, before you entered the educational system. Anyone who has experienced the intense questioning of small children and their insatiable and dangerous appetite for new experiences will know exactly what I mean. The continual inquisitiveness and curiosity, the determined acquisition of experience, the urgent wish to know more and minimal awareness of risk, so evident in young children, are what characterise the more successful graduates.

Most of the experience you acquire will hopefully be other people's, just as it was when you were a child. If you gain only your own experience, it

- tends to be painful, if not downright dangerous
- could lead to a loss of confidence
- can cost your employer a lot of money
- will certainly take far too long.

You would not expect children to learn about crossing the road purely by their own experience; the risks involved are far too high. Even as youngsters, they would soon realise (or be told or, even worse, learn from experience) the dangers. They would watch others doing it, ask for explanations and seek advice and then have their early attempts monitored closely.

To develop the skills and abilities you will need to fill more senior and responsible posts, you need to look at the decisions the holders of such positions take, particularly how and why they decide on certain courses of action. In this way, you will gain experience before you need it and, hopefully, make fewer mistakes when you succeed them. As an example, a new technician had produced many drawings of bolted steelwork connections, when the project manager decided to use welded connections. Three months' work was aborted and the technician made his displeasure clear! His mentor took him to one side: 'I want from you a discussion [a report] of whether that decision could have been made sooner, why it wasn't, and how you would have communicated the decision to the workforce.' Very early in his career, a technician was being advised to learn about the role of a project manager.

There is nothing new about experiential learning. At the dedication of New College, London in 1794, the scientist Joseph Priestley said,

> Whatever be the qualifications of your tutors, your improvement must chiefly depend on yourselves. They cannot think or labour for you, they can only put you in the best way of thinking and labouring for yourselves. If, therefore, you get knowledge, you must acquire it by your own industry. You must form all conclusions, and all maxims, for yourselves, from premises and data collected and considered by yourselves.

Priestley went on to say, 'And it is the great object of [our educational institutions] to remove every bias the mind may be under, and to give the greatest scope for true freedom of thinking.' With the exponential growth of knowledge since then, that ideal is probably no longer possible in education. But, as the world adjusts to the consequences of climate change, the depletion of traditional energy supplies, the conservation of limited material resources and the international marketplace, freedom of thinking must lead to vision, innovation and creativity if the United Kingdom, indeed mankind itself, is to survive.

There is no doubt that ICE3001 Appendix A, particularly for Chartered status, does expect you to be prepared to challenge established best practice, to think about the fundamentals for yourself and develop the understanding, vision and courage that will be needed to drive the industry forward beyond its current achievements. Whether you will have yet been trusted by your employer to employ such skills does not matter; you must demonstrate that you could if the opportunity arises.

Initial Professional Development

Most graduates start with a fairly narrow view of what civil engineering involves, but you should broaden your understanding and gain the confidence to have, and offer, your own views and opinions on a wide variety of topics related to your field and civil engineering generally. This is one of the purposes of the Development Objectives: to expand your understanding well beyond the technical details of engineering.

Membership Guidance Notes 21 and 22 offer topic areas which suggest what may be the subjects for the written part of the Professional Review. In fact, they also provide useful indications of the breadth of understanding which will be expected of a candidate during the whole Review process. So relevant topic areas have been included as a part of the framework for each attribute in subsequent chapters.

It is most unlikely that any employer will have the resources to teach you. In any case, I do not believe such support would be particularly effective. You will have to search for the necessary experience; much of that experience is in the minds of other people, although some of it may have been written down in books or codes and standards. You are going to have to do this searching by observation, discussion and enquiry. Like young children, in your mind you will be continually asking, 'Why?' and then choosing the most appropriate person from whom to seek an answer. Rarely will they volunteer the answers, because to them (with their greater experience) there simply is no problem – the course of action is 'obvious'!

My book *Effective Training for Civil Engineers* (1994) devotes two chapters to the task of learning through experience and, according to feedback I have received, they contain useful advice and helpful tips. I do not reiterate that material in this book, but the next chapter explains why everyone learns in different ways and suggests the kind of questions you need to be asking in order to develop the skills of learning through experience.

Learning styles

Not many people are aware of their own learning style, until they undertake some conscious research to identify it. As a result, they perhaps do not realise what sort of 'learning opportunities' best suit them. It is important to identify your preferred learning style so that, with help from your mentors, you can select, as far as is practically and commercially possible, the most appropriate type of activities to undertake. More importantly, when you clearly understand how you learn best, you can seek out opportunities for yourself. Or, conversely, at least understand why you do not learn as easily in certain situations.

For example, an 'activist' does not read the instruction manual first – they want to take it out of the box, plug it in and use it. In civil engineering (in contrast with most toys!), the consequences of such an approach can pose life-threatening risks, by replicating mistakes which have probably previously been made, recorded (and their repetition avoided) by others. So, it is imperative that much of the experience from which you learn is not yours, but is gained previously by others. This is one reason why the industry has so many standards and codes, setting down what has previously been found to be best practice.

There are many questionnaires on the market which attempt to uncover learning styles, such as Honey and Mumford, linked to the Kölb cycle which I modified in *Effective*

Training for Civil Engineers. Another popular one is known as VAK (visual, auditory or kinaesthetic). Again, there are several tests to determine which learning method best suits you but, in essence, the three letters stand for

- visual – seeing and reading
- auditory – listening and speaking
- kinaesthetic – touching and doing

and you may already have some idea of which works best for you.

There is a website which it may well be worth spending a little time exploring, to identify and follow links to various pages on learning: www.businessballs.com. A brief excursion into such a site and its links will give you a good indication of how you personally absorb information and make sense of the world.

Initial Professional Development for Civil Engineers
ISBN 978-0-7277-4147-9

doi: 10.1680/ipdce.41479.021

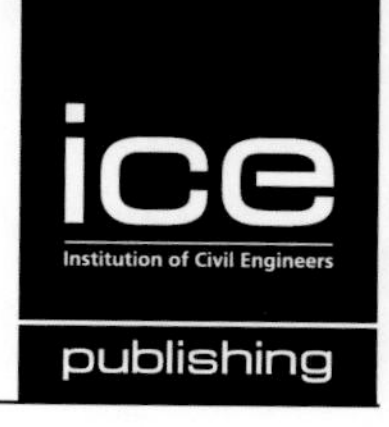

Chapter 4
Advice on learning

Most young engineers are usually able to record 'information' or facts, on a day-to-day basis. It is this rudimentary skill which they apply to their routine reports in a training situation. But this is only the first (and possibly the easiest) part of learning through experience. Experience is not only doing or seeing something and recording it – it is asking yourself what you learnt from it. Whenever you record an event, always (at least in your own mind) end your sentence with 'because...' and go on to give the reasons for the outcome (which may require further investigation, probably through discussion). Every statement must include 'How' or 'Why' or the other 'honest serving men' (see the quote from Rudyard Kipling in the Acknowledgements). In this way you develop the mental agility to draw out the learning from the experience: standing back, looking at what you did/saw/heard and asking 'What did I learn from that?'.

When you first start out on your IPD, nearly everything is new, so there is the potential to learn something from almost everything that happens. But you cannot record every single thing, every conversation that goes on around you. So how do you sort out what is worth recording? Try to distinguish the elements that are most relevant to you at that particular point in your development. This is where the Development Objectives can help as a broad framework within which to organise relevant experience.

For example, it is likely that you will start your IPD by extending from your education. So you may begin by gaining experience relating to attributes 1, 2 and, almost certainly, 6. As you progress, the personal attributes 4, 8 and 9 will become increasingly important. Following on close behind will be a greater involvement in the environmental and commercial aspects of the business, attributes 5 and 7. There is no certainty that your progression will follow that route, but it is probable.

Many young engineers appear unaware of the experience available to them – it seems to pass them by. An example might help to explain this. I asked a trainee on site what he'd learned that week. He could not think of anything. So I asked him to tell me about what had happened during that week. One of the 'events' he listed was that there had been a flood and 'the water nearly got into some of the houses over there...'. What a gift! But he didn't recognise the potential that this occurrence held for him. Why had the flood happened? Could it have been avoided? How was inundation of the houses avoided?

Were there any other steps which, with hindsight, could have been taken to avoid or at least mitigate the risks of flooding? So much learning to be gained through one small experience!

Another problem for most engineers is remembering what the problem was. Once we have solved problems, they become invisible. It is as if they never existed, because they are consigned to the 'deleted' box in the brain. Candidates for Review often tell me that on their project there were no problems. There most certainly were, but by that time they had gathered sufficient experience to know what solutions were possible, and how to choose between them, so there were no problems *for them.*

The key is to hold on to the problem, stand back and recognise what it will enable you to demonstrate. Ask yourself a series of generic questions.

- Did anything unexpected happen?
- Did we find anything unexpected?
  - If so, what implications/consequences were there?
  - If not, what could/might have happened or be unexpected?
- Was this 'the only way' things could have been done?
  - What other options were discarded and why?
  - Why was it decided to do it this way?
  - Advantages/disadvantages of the chosen option?
  - Was it a compromise?
- What were the anticipated outcomes?
  - What were the possible consequences, implications and barriers?
  - Were any short-/long-term future benefits or problems identified?
  - What actually resulted?
- With hindsight, would you do the same thing next time?
  - If not, what would you do and why?
- Back to options, decisions and outcomes.

Two questions you must ask, throughout your IPD (and, indeed, beyond) are 'How did I influence the progress of the project?' and 'Was the outcome better than it might have been without my input? And, if so, in what respect?'.

Remembering what the problem was is not only a difficulty for the 'novice' engineer: it applies equally to the person who is approaching Professional Review: where to start preparing the submission? It is a different problem but equally challenging. By that stage, you will have achieved so much and done so many things, many of which will have become routine. The problem starts with the question 'What do I select from this vast resource of "experience" to demonstrate my attributes . . . ?'. The earlier you begin to consciously learn through experience, the greater will be the learning outcome, and the easier and quicker it will be to develop the required attributes. Recording this learning means that you will be able to readily demonstrate your abilities to the Reviewers. Importantly, you will have learnt to avoid clichés such as 'I managed the office resources' which, to a Reviewer who is trying to get to know you, could merely mean 'I put the kettle on'.

Detailed techniques for learning through experience

The questions outlined above perhaps need more detail, particularly when you are a novice just beginning to learn through experience.

Learning from experience is not necessarily actively participating, but could be watching what is done and working out, by questions, discussion, research and thought

- what happened
- why it happened
- what alternatives were rejected
- why it was decided to follow that particular course of action.

This cycle of questions is the method by which you can gain maximum benefit from even limited experience. Where you are required to write regular reports as part of your formal training, answers to such questions should form the content. If there is no formal requirement, it is still a good idea to record your thoughts on what has happened; these notes will be most useful as you recall experiences later.

In all the work in which you participate, judgements are being made and decisions taken on all manner of things. Senior or more experienced engineers are making judgements based on

- their experience (having encountered the same circumstances previously)
- instinct (a mental agility based on an acquired familiarity with similar situations)
- imagination (the ability to predict what might happen, given this particular set of circumstances).

What you must do is find out what those judgements were and how and why they were made. You need to explore them, discuss and think about them to arrive at your own opinion of their validity. This process is the major reason why the ICE strongly recommends that you write regular reports, around which you can build conversations with your mentors – your Supervising Civil Engineer or Delegated Engineer and anyone else whom you think can help. They all have great experience from which you can learn. Sooner or later, as you move up the hierarchy, you will have to make similar judgements and, by then, you will have amassed a significant bank of experience on which to base your decisions.

These decisions inevitably vary for every project, from materials research to maintenance, from design to demolition, from defining the problem to identifying the most appropriate solution. The decisions being made require answers to a whole series of interrelated questions. What follows are suggestions (neither exclusive nor inclusive) of the types of questions these might be.

Technical understanding

- What are realistic loadings (dead and live weights, forces, flows) for the geographic, social and environmental situation and for the foreseeable usage?
- What adequate factors of safety should be applied?
- What is the most realistic method of analysis?
- How can I satisfy myself that the results are realistic?

Technical application

- Is the proposal realistic and the best technical option ('appropriate technology')?
- Can the proposed solution be implemented safely, economically and realistically?
- Are the necessary resources (the 'five Ms' – manpower, machinery, materials, methods and, of course, money) available, sustainable and can they be delivered on schedule?

Financial

- Is the proposal affordable?
- Where will the money come from?
- Does the proposal provide value for money for all parties affected?
- What are the longer-term costs of, or income from, the proposal?
- Are there any unlikely expenses which might arise from the work and would contingency funding become available (e.g. insurances)?

Environmental

- What are the effects during and after the project?
- Do the positive gains outweigh the detrimental effects?
- Is the proposal sustainable, in both the short and long term?
- Has due consideration been given to alternatives?
- What are the likely challenges or special interests?
- Does the proposal comply with our/local/national/international policies and protocols?

Health, safety and welfare

- What are the hazards inherent during and after the project?
- How can these hazards best be eliminated or, at least, mitigated?
- Can the proposal be achieved, used and maintained:
  - within the safety and health standards pertaining in the vicinity,

  - without detriment to all affected, locally, nationally and internationally (e.g. carbon emissions, air and water pollution)?

Community

- What will be the impact on the locale before, during and after the proposal (businesses, residences, transport network)?
- What will be the impact on all those affected by extraction or manufacture of materials and their transport and use?
- Will any ethnic communities (implications for holidays, religious observance and festivals, language) or minority groups (e.g. work programmes, special needs, disabilities) be affected?

Commercial

- Is this the most economic way to solve the problem?
- Are there any false economies and could they be eliminated?
- Can a profit be made or will the project, at least, break even if the problem is solved in this way?
- What are the financial risks and how are they mitigated?
- Which group carries each commercial risk (client, contractors, consultants, suppliers, end users)?
- What could go wrong unexpectedly and what would be the effect on each party involved?

Contractual

- What forms of contract are there between us and our client and between other parties to the project?
- How do the contracts apportion the major risks?
  - Have these risks been 'hedged'? If so, how?
- Are standard forms of contract being used? If not, what were the modifications and why were they made?

Political

- How does this project fit our business?
- How are we going to successfully present this project to the authorities/public?
- How is the proposal going to be received by local politicians and their constituents?
- Does the proposal comply with local and central government policies, statutes and regulations?
- Does the project comply with European Union (EU) policies and regulations, if applicable?

Organisational

- How does this project fit into our business plan?
- Is this work within the defined context and direction of the organisation?

- Do we have all the necessary resources? If not, can we obtain them?

Very few, if any, of these decisions will be made by you at this early stage in your career, but most of them will become your decisions as you take on greater responsibilities. Repeatedly asking about such matters, and finding and recording the answers, will quickly develop your own capacity to make judgements, at first probably more technical but soon more holistic.

Initial Professional Development for Civil Engineers
ISBN 978-0-7277-4147-9

doi: 10.1680/ipdce.41479.027

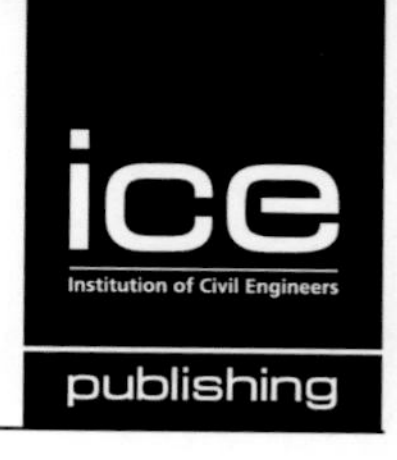

Chapter 5
Technical competence

It is reasonable to suppose that, having spent a considerable part of your academic life studying the principles of structures, hydraulics, soil mechanics, materials and the like, your technical competence would already be well-formed. The same applies if you have a degree in related sciences and are progressing towards Associate Membership. However, statistics from the ICE Reviews suggest otherwise, with attributes 1 and 2 regularly being one of the causes of rejection.

Institution framework

The Institution offers the following descriptors for this attribute in Appendix A of ICE3001:

1. Engineering Knowledge and Understanding
Maintain and extend a sound theoretical approach to the application of technology in engineering practice
Use a sound evidence-based approach to problem solving and contribute to continuous improvement

and additionally for Chartered status:

Maintain and extend a sound theoretical approach in enabling the introduction and exploitation of new and advancing technology
Engage in the creative and innovative development of engineering technology and continuous improvement systems

2. Engineering Application
Identify, review and select techniques, procedures and methods to undertake engineering tasks
Contribute to the design and development of engineering solutions
Implement design solutions and contribute to their evaluation

and additionally for Chartered status:

Conduct appropriate research, and undertake design and development of engineering solutions
Implement design solutions and evaluate their effectiveness

There are comparable attributes for Associate Membership (AMICE) where, essentially, 'technical' or 'scientific' replace the word 'engineering'.

Membership Guidance Note 47 gives advice and guidance specific to engineering knowledge and understanding. It states that the Reviewers 'will be looking to discover if you are able to understand the behaviour of the physical world you encounter and have the ability to visualise this world in terms of loads, forces, deflections and movements'.

Engineering knowledge

Your technical competence is wholly dependent on the field in which you operate (e.g. structures, highways, flood defence, geotechnics, environmental management or economics). *ICE3001 Routes to Membership* states at the start of Appendix A that 'the interpretation of attributes 1 and 2 will relate to your fields of work... You must demonstrate a sound understanding of core engineering [technical or scientific] principles in those fields'. Thomas Telford Ltd and others have a number of publications and courses relevant to many specialisms, well beyond the scope of this book.

The core engineering principles do not change with time. They are mathematical or empirical formulae, sometimes laws, defining, for example, the way a beam or a frame bends, the way a liquid flows down a pipe, the way heat passes through a wall or water permeates the ground, which are independent of the materials themselves. Many of them are known by the name of the person accredited with defining them, such as Bernoulli, Darcy, Euler, Young, Chézy, Newton and Hooke.

It is these principles which enable you to demonstrate that you understand what you are doing; that you are not solely reliant on codes and standards, colleagues or software, but have the core knowledge to be able to discuss alternatives and options in principle, before committing to detailed analysis. An understanding of these principles is also vital in checking that the answers obtained from any software programs seem sensible.

These are the principles that you initially learnt as facts and practised applying them to well-defined analytical problems, perhaps with little understanding at that early stage. In the workplace the formulae and procedures will begin to make more sense as you use them to resolve real and probably ill-defined problems. What you must avoid is any subconscious complacency when you are called upon to utilise defined and often well-established systems. It seems that many younger engineers are rushing into the use of some highly sophisticated programs, apparently understanding neither the engineering principles upon which the software is based, nor what assumptions have been made in the program or in adapting the problem to fit the program, nor, indeed, why that particular software was chosen. This is a dangerous tendency.

The reliability of a computer's output is entirely dependent on the quality of the information it is fed, and that quality is dependent on a fundamental understanding of the basis on which the program works. Finite element analysis, for example, is an incredibly useful tool. However, if you do not fully understand the principles upon which it works, then you may apply it incorrectly and, no matter how good the system, the results will not be reliable.

Neither general design principles nor sophisticated analytical techniques are able to offer either the right answers to specific problems or give prescriptions for good design. Most of established best practice, as defined by codes and standards and taught in universities, offers formulaic answers and design mechanisms which were originally created by inspiration, repeated experimentation or mathematical representation. These formulae have been found to repeatedly offer realistic solutions to analytical problems and so, quite correctly, have been promoted as best practice. But they may not be correct for all problems faced today.

Textbooks too often, perhaps by default, present theorems and proofs as seemingly perfect and inviolable, without explaining the hunches, insight and approximations which were explored to derive them. As an example, Henry Darcy derived his law for the movement of water through the ground by experimenting with sand beds. Extrapolating into other materials could therefore invalidate the theory. Sometimes subjective modifiers can be applied to overcome any discrepancies, but you have got to decide whether the results are representative or not. Leonard Euler, a mathematician, derived a theorem for the bending of struts which has been found useful within certain limits (derived by experience and experiment). Other theorems, such as those derived by Rankine–Gordon and Perry–Robertson, have since been found to be more representative at lower slenderness ratios.

Formulae are not applicable to every situation, and you must understand them in principle in order to decide whether they are truly applicable to your problem, whether you should use any results merely as a guide, or whether you need to consider other methods.

Design is largely a creative process, based on established methods, techniques and frameworks which structure and guide the designer, who must understand the basic engineering principles related to stability, form, strength and effectiveness. Previous best practice will not necessarily always solve a problem, but it most certainly will offer the person using it a sound basis on which to decide whether it is applicable or not. This is vitally important where engineers, for reasons such as economics and aesthetics, are routinely approaching the limits of the strengths of materials or functional capability.

Ancient cathedrals, civic buildings and bridges were so heavy that applied loads caused only a small part of the stresses and strains in the structure; what we call 'built-in redundancy'. Nowadays, external forces are the major creators of stress, because the structures are relatively light. And those external loads are perhaps less predictable than in the past. In the Manchester bombing of 1998, for example, massive older buildings survived quite well, but several newer ones had to be totally replaced, because the external forces were greater than anticipated during their design. Perhaps we should be considering building more redundancies back into our structures. Electronic software, formulae and standards, as well as routine procedures, can only provide a framework within which design can be rationalised but by which it must not be constrained.

Engineering understanding

My computer dictionary contains a simple definition of understanding: 'Able to know something; able to comprehend the sense or meaning of something.'

Understanding is the culmination of a thought process where you

- acquire knowledge (about 'something')
- apply the knowledge to an appropriate situation (for example, a problem, set of circumstances, issue)
- identify the options (the brain 'searches' itself and pulls out all the times you've looked at this problem before and identifies similarities and differences, pros and cons, advantages and disadvantages, did it work last time?)
- arrive at a decision (a judgement based on the brain having processed all the above information) and are 99% sure you are correct for this particular circumstance.

There is much more about 'understanding' in Chapter 2.

Engineering application

It is vital when applying technology in engineering practice that time is spent understanding in principle how that particular piece of infrastructure is going to behave (e.g. how the building frame is going to deflect under a variety of loading conditions, how quickly heavy rain will be dispersed into the watercourses, what parameters are going to affect the forecast of future flows – water, people, traffic). In implementing designs or maintaining infrastructure, you must understand how they are going to perform in an incomplete state. For feasibility studies, you must know in principle whether the options being considered are realistic solutions. Any engineer must be able to answer questions such as

- What suitable methods are available to me?
- Which is the most appropriate?
- Which software program (if any) fits that decision?
- What assumptions are inherent in the program?
- Are those assumptions valid for my problem?

- What assumptions do I need to make to make my problem fit the program?
- How can I validate my assumptions?
- How do I decide what are the critical cases to be analysed?

And, perhaps the most important of all

- How do I satisfy myself that the results I get are realistic?

This last question requires you to be able to do some 'back of an envelope' calculations to convince yourself that the answers are in the right order of magnitude and look realistic. Such calculations might best demonstrate at Review your full understanding of the engineering principles, since you would, of necessity, have to simplify the problem down to its basics.

For example, you might be undertaking the analysis of a pipe network and spot an inconsistency in the output. A CCTV survey could subsequently identify a pipe run which you did not anticipate in the input data. How did you detect the inconsistency? Similarly, you might do a simple analysis of a roof truss, reduce a building to a rectangular two-dimensional portal or do a simple analysis of one side of a cofferdam. When a road was required with a design life of ten years, manuals and standards did not provide answers and I had to revert to first principles.

So often, experienced engineers say something like, 'That result is not what I expected'. How did they know what to expect? To be able to develop realistic expectations requires a sound understanding of the principles underlying the problem. Perhaps these simple checks *are* done, but Review candidates do not fully realise their importance in demonstrating an understanding of technical principles and so omit to tell the Reviewers about them.

All of the above is relevant to anyone advancing to ICE Membership, whether recognised by the Engineering Council as IEng or CEng, but for potential Chartered Engineers (CEng), an added requirement is to enable 'the introduction and exploitation of new and advancing technology', so you must demonstrate that you are fully aware of new developments in your particular area of expertise and, wherever possible, seek to utilise them. This means keeping up to date with developments by reading technical journals and advertising literature relating to your field and, whenever possible, attending exhibitions and demonstrations of new technology.

You are further required to 'engage in the creative and innovative development of technology'. So you must be able to offer examples of original thinking, where you perhaps thought of a novel or ingenious solution to a problem and persuaded others of its validity. It may never have been used due to constraints beyond your control, but that does not matter. You must demonstrate that you are continuously looking to improve systems, methods and outcomes, and are not comfortably satisfied with the status quo.

As a Chartered Engineer, you must realise that yesterday's answers will probably not be appropriate to solve tomorrow's problems; you must demonstrate that, given the chance, you are capable of leading teams in tackling new challenges. At the time of your Professional Review, you may not yet have had those chances, but you must show that you could rise to the challenge. So there is an element of expectation to the Review process. Obviously, the more experience you gain, the more opportunities you will have; so the later your Review, the more your capabilities should have been utilised.

Initial Professional Development for Civil Engineers
ISBN 978-0-7277-4147-9

doi: 10.1680/ipdce.41479.033

Chapter 6
Management and leadership

There is much confusion about the relationship between leadership and management. At one level, management *is* leadership but, very often, management is the administration of established systems and routines. So it is clear that, although they are not mutually exclusive, they are not the same. The Member Professional Review (MPR) is more management orientated, while the Chartered Professional Review (CPR) is very clearly looking for leadership qualities (see the next section). It is important that you understand the differences and similarities so that you can make an informed decision as to whether MPR or CPR attributes are more appropriate when you present yourself as a potential Member and target your preparation accordingly.

Institution framework

The Institution offers the following descriptors for this attribute in Appendix A of *ICE3001 Routes to Membership*:

Plan for effective project implementation
Manage the planning and organisation of tasks, people and resources
Manage teams and develop staff to meet changing technical and managerial needs
Manage quality processes

In addition, for Chartered Engineers:

Plan, direct and control tasks, people and resources
Lead teams and develop staff to meet changing technical and managerial needs
Commit to **continuous improvement** through quality management:

- quality plans and systems
- maintain quality standards
- quality records
- contribute to improvement of quality systems

The Development Objectives offer:

C1 Plan organise and supervise resources to complete tasks
Plan for effective project implementation:
- Planning and resourcing, programming
- Method statements
- Information and administration systems
- Instructions
- Records

C3 Develop people to meet changing technical and managerial needs
- Lead by example

C4 Continuous improvement through quality management
- Quality plans and systems
- Maintain quality standards
- Quality records
- Contribute to improvement of quality systems

Membership Guidance Note (MGN) 22 adds *team building* for Membership, while MGN 21 for Chartered Membership adds *team leadership*.

Overview

Most of us have an amalgam of skills and attitudes which rarely, if ever, fit logically into defined boxes. There do, however, seem to be basic differences between a manager and a leader, perhaps best set out in a comparative table:

Manager	Leader
Administers	Innovates
relies on established practice	looks to be original
Maintains	Develops
focus on systems and structure	focus on people
Controls	Trusts
short-range targets	long-term goals
eye on bottom line	eye on the horizon
Imitates	Originates
accepts status quo	challenges status quo
'does things right'	'does the right things'

Leadership

Leadership is an intangible quality which most people can recognise but no one seems able to define. When people are asked to name prominent leaders, the one thing that distinguishes any list is its diversity, from Hitler to Mother Theresa, from Nelson Mandela to Margaret Thatcher. So there must be many styles of leadership and no one clear definition. Like management, there are literally hundreds of attempts to describe leadership. 'Lead' derives from a word common to old north European languages, meaning a road, a path, the course of a ship, so perhaps the term has something to do with setting a direction and movement along it.

The traditional view (which still manifests itself in some cultures and circumstances) is a fundamental belief that the leader is right and that coercion and physical rewards are needed to persuade people to put the leader's 'correct' decisions into practice. When the leader is largely perceived to be wrong, however, hierarchical power or force can only motivate people for a limited period. So true leadership is not executive power.

David Gilbert-Smith, founder of the Leadership Trust, said,

> True leadership comes not from rank or status but from personal power. Leadership means winning people's hearts and minds. To do so requires first winning one's own. So leadership development begins with the confidence of self-knowledge and the calm of self-control. These are qualities which can be developed, not easily, not by mechanical application of any 'leadership theory', but by looking long, hard, truthfully at our behaviour and our effect on others and then – crucially – committing oneself to the personal responsibility for doing something about it.

In discussions with Reviewers, one told me that he was scared by excessive self-confidence, but was certainly seeking 'self-belief' in candidates. The difference is, in essence, the contrast between immediately 'knowing' what the answer will be (jumping to conclusions) and having the quiet confidence in your own ability to seek and find the best answer, utilising the knowledge and experience of the whole team.

Clare Short in her book *An Honourable Deception?* (2004) about recent governments, states

> Good leadership draws people together in a shared endeavour. It forges agreement on the best way forward and then decentralises power and creates structures that encourage all to contribute to the task in hand. It is this kind of leadership that achieves large advances because highly motivated people working together for a common purpose create highly effective organisations.

The way in which people are led is now central to the success, indeed the survival, of every enterprise. If people at every level of an organisation are not encouraged to use their initiative, do not want to contribute to the limit of their potential and do not want to learn and share their experience within a team, then momentum will be lost and the organisation will fall behind.

Differences between a Member and a Chartered Member

The centre column of Appendix A of *ICE3001 Routes to Membership* uses words such as 'application', 'plan', 'prepare', 'sound knowledge' and, significantly, 'manage'. Potential Member must demonstrate that they can manage their own contribution to the project and 'organise tasks, people and resources', perhaps more the role of a manager. This is in addition to the engineering capability of 'solving problems' by 'applying technology' with an 'evidence-based approach', using 'techniques, procedures and methods' to develop solutions – the abilities which distinguish an engineering manager from any other type of manager.

Many civil engineers work in complex organisations which have developed systems to 'organise tasks, people and resources' and now have in place written procedures. In such circumstances, it is necessary, of course, for engineers to comply with those company procedures, but this does not preclude consideration of whether they think such systems are always the most appropriate course. In other words, they should not merely unquestioningly accept them, but should think about why they are in that form. For example, if you work for local government, why are there Standing Orders and how do they work? What is their purpose in regulating the proper behaviour of elected members and staff? Are they successful?

To demonstrate that they fully understand how to manage their own contribution, candidates for Review should perhaps have thought about and recorded how, if they were ever given the chance, they might improve such procedures. Perhaps, as a qualified professional engineer, you may be called upon to do so!

The right-hand column of Appendix A uses many words, such as 'direct and control', 'enable', 'exploit' and, particularly, 'leading', which show that potential Chartered engineers must demonstrate that they are not only competent, but have the capability to drive the business and their colleagues forward, with greater efficiency and more effectively. This ability to lead is not a function of your job or position in the hierarchy, it is more an attitude of mind: a dissatisfaction with the 'norm', a desire to do things better and the ability to convince and motivate others to support you.

Many, if not all, of the projects with which you become involved, will need an increasingly sceptical public to be persuaded that the proposed solution is the best answer to the problem. This will require you and/or your colleagues to explain the proposals in a public arena and to counter attacks, sometimes quite vitriolic, from those who perceive that they are being disadvantaged. In such circumstances, you will have to demonstrate self-belief, integrity, positive arguments and overwhelming competence to persuade the public that you are correct – most, if not all, of the qualities of a successful leader.

It is most important for your success at a Review to realise and understand this basic difference between the two classes of Member. It is quite possible that, at the time you seek professional qualification, your personal development is somewhere between the two classes. If so, then choose the appropriate Review *at that time* and make sure that your preparation, submission and performance will demonstrate the appropriate abilities. In practice, there are relatively few people who develop into outstanding leaders. Many, however, will eventually be capable of motivating and managing successful teams. To paraphrase Sir John Blashford-Snell, the explorer, many people have a spark of what it takes to lead. Leadership is there – it is a matter of allowing people to find it and use it. So, it may take time, and the Institution's routes to membership recognise this.

You may, of course, believe you are already capable of becoming Chartered. It may well be that, at the time you apply, you are frustrated because you are not yet in a position which gives you opportunities to change things. But this does not preclude you from demonstrating that you could, given the chance. You must demonstrate that you are thinking fundamentally about the 'use of scarce resources, care for the environment and public health and safety' (Royal Charter) and that you have ideas on how things could be done better.

Quality processes and management

What is quality? It is not necessarily 'the best', 'most expensive' or 'highest quality'. Products with low prices can be considered quality items if they meet a market need ('value for money'). The standard of quality is determined by the intended users, clients or customers rather than a uniform standard set by society itself. But, at the same time, the production of that product – its raw materials and components, the management of the methods and processes by which it is produced, are also part of the process of achieving quality.

So, perhaps a quality project should be suitable for the intended purpose ('fitness for purpose') and mistakes in its procurement should be eliminated ('right first time'). Any programme for the systematic monitoring and evaluation of the various aspects of the procurement of a project, service or facility which tries to ensure that standards of quality are being met is called quality assurance (QA). QA cannot, unfortunately, absolutely assure or guarantee the production of *quality*, but it does make this more likely.

Quality control (QC) is the testing of the quality of aspects of a project, product, service or facility once they have been produced, to make sure they conform to specific requirements and comply with predetermined standards. It should be used to pick up any deficiencies in the QA processes. The presence of any defects can then be used to identify and rectify deficiencies in the process; so QC should be an inherent and iterative part of QA.

Most organisations have systems and procedures for efficient and effective quality management of a project through every key stage:

A	Commit to invest
B	Commit to implement
C	Available for use
D	End of defect liability period
E	End of lifetime of project.

Between these are the processes, in some of which you will be involved:

Before A	Feasibility studies, sourcing finance and identifying resource availability
A to B	Planning, design and approvals
B to C	Construct or implement
C to D	Defect liability period and use
D to E	Use, maintenance and modification.

Between C and E is the obvious period in which quality issues are highlighted. But they are present at every stage, it is just more difficult to identify them. Errors of judgement in A and B can have profound effects on the resulting quality of the preferred solution, and great care must be taken to ensure that the best option is identified and procured.

A quality issue is any issue that affects the project so that work has to be redone, modified or a compromise made at a lower standard than originally agreed. So issues include late changes in client expectations and budgetary and environmental limitations on specifications, as well as the remedying of errors and mistakes. The elimination of any of these elements markedly improves efficiency, delivering what is required at reduced cost. Every engineer must know how their organisation manages these issues and collect examples where changes or mistakes made quality more elusive. Make notes of any factors that compromised the quality of your contribution and think about how they might have been avoided.

Clients want projects delivered by profitable companies

- on time
- within budget
- free from defects
- safely
- efficiently (right first time)

but they do not always know exactly what they want or need after stage A.

It is therefore vital that anyone involved in the planning and design of projects attempts to clarify and fix the client's brief as early as possible in the procurement process, if quality is not to be compromised. Unfortunately, many candidates for Review will have experienced the frustration of late changes or uncertainties in the client's expectations. Reviewers will expect candidates to have learnt by the hindsight gained and to have ideas on how such problems may be avoided in future.

To achieve better performance through continuous improvement, John Egan, in his report *Rethinking Construction* (1998), presented to the UK Government in 1998, stated that there must be realistic but extending targets. This is why you have probably been benchmarked against Key Performance Indicators (KPIs) or similar, of which Egan identified seven main groups

- time
- cost
- quality
- changes
- client satisfaction
- business performance
- Health and Safety.

Significant improvements have been made since then, but they have not reached the targets originally set by Egan's report. One area of significant change has been an increase in partnering in the supply chain. Longer-term relationships (not frequent competitive tendering) with clear measurements of performance against KPIs, utilise specific expertise more effectively, and create mutual respect and trust, which in turn bring a sustained improvement in quality and efficiency.

The state of the UK economy at the time of writing (2010) is such that clients are said to be seeking a further 20% reduction in budgeted costs ('more for less') beyond what has already been achieved. How is your organisation facing up to such challenges? Which of the five key drivers of change identified by Egan, and listed below, are being utilised to drive further improvements and how are they being interpreted

- committed leadership
- customer focus
- integrated processes and teams
- quality driven
- commitment to people.

Find out about (and make notes on) recent changes in (and future proposals for) improvements in safe and decent working conditions, and what efforts are being made at every level to continuously improve management and supervisory skills.

Are you part of an Early Contractor Involvement process? Whether you are or not, what efforts are you aware of to design for ease of construction, by fully understanding the true costs of temporary works or the difficulties of coping with unpredictable ground and weather.

Is your organisation utilising repetition, standard components and routine processes in an attempt to develop familiarity and prevent mistakes?

The public sector, as the major UK client, has a vital role in leading the development of a more sophisticated and demanding customer base. Some progress has been made, but what further improvements can you envisage?

Initial Professional Development for Civil Engineers
ISBN 978-0-7277-4147-9

doi: 10.1680/ipdce.41479.041

Chapter 7
Independent judgement and responsibility

Adequate demonstration of independent judgement seems to cause too many candidates difficulties at Review. Again, this could well be a legacy of their academic education, where they were expected to always give the 'right' answer. In the Review, however, they are expected to show examples where they did not fully comply with established best practice, but made judgements (probably in discussion with other more experienced engineers) about the best practicable solution to that particular problem. In practice, most of us are forced by circumstances to make these judgements all the time, but it appears that many Review candidates seem to think (mistakenly) that they must demonstrate that they always comply with the 'rules'.

Institution framework

The Institution offers the following descriptors for this attribute in Appendix A of ICE3001:

> Identify the limits of personal knowledge and skills
> Exercise sound independent engineering judgement and take responsibility

and additionally for Chartered status:

> Identify the limits of a **team's** skill and knowledge
> Exercise sound **holistic independent judgement** and take responsibility

Additionally, the Development Objectives offer:

> **C1 Develop people to meet changing technical and managerial needs**
> - Objectives and work plans
> - Support and help others to develop

None of the topic areas outlined by the MGNs for the Written Test or Assignment for Membership specifically mentions these qualities, but many would require their use to underpin the content of the answers.

Independent engineering judgement

Engineering judgement is the ability to compromise.

- How do we decide between what forces or loadings any structure should be designed to withstand and what conditions are unrealistic?
- How do we determine what loads the ground will sensibly support?
- How is the requisite capacity of a pipe or a road decided and the necessary flow rate achieved?

Of course, we rely on established best practice, as defined in codes and standards, but these do not always reflect the circumstances particular to our problem. Sometimes a formal departure or relaxation has to be sought. In other cases we make a judgement that, while the result does not quite comply with established best practice, in our considered opinion it will suffice. Always remember, you may have to justify your decision later, so the documentary evidence you might need for your Review should always be there in the quality system, even where a formal agreement to the decision might not exist.

Virtually all potential Members probably make such judgements frequently even if it is merely, for example, rationalising the spacing of reinforcement bars from that calculated, slightly amending the traffic sight lines or cross-over distances, or allowing the pipe surcharge in extreme flow conditions to exceed the optimum. What it appears they fail to do is to demonstrate this ability during their Professional Review. Perhaps this is a residue of the many examinations they have passed, where their answers had largely to be 'correct'? So candidates may be unwilling to give examples where they 'broke the rules' and hence fail to demonstrate engineering judgement.

Holistic independent judgement

Many of the judgements you will make early in your career will be associated predominantly with technical matters. Generally, these are risks taken by Members and Associates, usually within fairly strict guidelines and often after discussion with more experienced personnel. But many such judgements are becoming more taxing, requiring an overview of many considerations well beyond the purely technical – environmental, economic, social, global. These judgements are mainly the province of Chartered Engineers and evidence of adequate understanding of how to balance these conflicting considerations must be demonstrated at the Chartered Professional Review. For example

- the consequences of climate change and the depletion of traditional energy sources are forcing us to rethink issues such as flooding risk, coastline protection, insulation and modes of transport
- to achieve more sustainable development, traditional high specifications are being challenged to enable the use of secondary raw materials.

Further, public perception is continuously changing, and the pressure on Chartered Engineers to react positively is increasing. So many factors have to be taken into account when deciding on the best course of action. 'At this time, and for the foreseeable future, with these resources, in these circumstances, this is the best compromise' is a very demanding holistic decision, and cannot rely purely on the decisions taken yesterday, as defined by codes and standards.

Change is endemic and becoming ever-more rapid, not only with regard to 'circumstances' and 'resources' but also the future. For example, established best practice in a developed country may not offer the most appropriate solution when it is transferred to a less well-developed one, a problem identified by the expression 'appropriate technology'. There are unfortunate examples, well-documented, of sophisticated sewage works which were never operated and European-style motorways where the hard shoulder and inner lane are used by camel trains and donkey carts. In the UK, there are trunk roads where many villages were bypassed in the 1960s by dual carriageways. Is it appropriate to design further bypasses on the same road to modern standards, or to reproduce the 1960s standards for consistency throughout that stretch of road?

Codes and standards are written on evidence from the past from specific societies, and our industry is being required to anticipate the problems of the future all over the world. For the Chartered Professional Review, it will be necessary to record examples of occasions where you did not slavishly follow established best practice, where you made a judgement and decided, probably with discussion at this stage in your career, that what you were proposing was the best solution in the circumstances. From such examples you can demonstrate your capability to 'exercise sound holistic independent judgement and take responsibility'.

It is unlikely that, at this early stage in your professional career, you will have had (m)any opportunities to make these decisions yourself. You must demonstrate that, when any opportunity arises, you have the capability to grasp it, having developed a wide perspective and having the necessary intellectual determination. Not necessarily that you *have* done it, but that you *could*.

Identify limits of knowledge and skill

One of the greatest responsibilities of all engineers is 'knowing when they don't know'. It is all too easy, for example

- to rush into resolving a problem without realising all the implications
- to use a piece of software without checking that the assumptions in the program are valid for your problem
- to cursorily thrust advice on a client without a thorough knowledge of their problems.

You must record occasions on which you decided to seek advice and guidance from others, or greater clarification, and how and why you reached that decision. We often tend to leave it just a little too late, forlornly hoping that the answer will suddenly materialise. Develop confidence in your own abilities, but also feel confident about seeking support and guidance whenever you feel insecure.

For potential Chartered Engineers, you are required not only to be able to identify your own shortcomings but also 'identify the limits of a **team's** skill and knowledge'. Do not take this to necessarily mean only subordinate staff or those working for the same employer. It is important that you judge the capability of all those with whom you work, otherwise you

- will not be able to communicate with them accurately or effectively
- may ask or expect too much of them, which could be demoralising
- might ask too little of them, which will be demeaning
- might have too much/too little faith in their judgement, to the detriment of the advice and support you expect.

Develop people

From an early stage in your progression to Chartered membership, you will be 'using' people to provide information, expertise and support. As explained in the previous section, this will require you to assess their capability. But, to gain maximum cooperation and the greatest enthusiasm, you will need to imbue them with a feeling that they are providing something useful and helpful – with a sense of worth.

If you are to lead and develop them to the best of their abilities, you must

- ensure cost-effective use of their existing skills and abilities
- create opportunities whenever possible that are mutually beneficial and rewarding
- make them feel secure in tackling new, more demanding roles
- provide them with support without detracting in any way from their own sense of responsibility
- continuously challenge them to achieve more
- offer praise and positive criticism as appropriate.

All of these requirements are dependent on mutual respect and trust. You will see that these aims are very similar to those listed in MGN 12 as the requirements for a Supervising Civil Engineer, who is, of course, trying to lead and develop trainees to professional competence. Developing relationships and making your colleagues feel needed and valued (building teams, whether within the same employing organisation or not) is one of the characteristics that single out potential leaders.

Summary

The requirement for Chartered Civil Engineers, in particular, to show that they are capable of original thought and determined innovation has always been part of the Review criteria, subsumed (in the predecessor 2000 series) under 'Vision and Leadership'. But it has never before been spelt out as it is now.

The 3000 series leaves no doubt at all that, particularly for Chartered Engineers, competence on its own is no longer enough. The descriptors for competence and application require you to demonstrate that you have the vision and determination, underpinned by a fundamental understanding (see Chapter 2) to enable you to improve on established best practice, and to move the profession and the business forward to solve the intangible problems of the future.

This does not necessarily mean that you have to have done something earth-shattering before your Review, but you must demonstrate that, when the chance or opportunity becomes available, you have the self-belief and capability to take full advantage. One of the questions the Reviewers consider is, 'Would this person, when in a position of responsibility, make the correct decisions?' Note the importance of the wording: 'Would' not 'Has'. The Reviewers will be assessing your capability for the future, not your competence in the past.

Initial Professional Development for Civil Engineers
ISBN 978-0-7277-4147-9

doi: 10.1680/ipdce.41479.047

Chapter 8 Commercial

The acquisition, use and control of finance are central to every civil engineering enterprise. All of the resources used by engineers (manpower, machinery, materials, methods) are dependent on the fifth (the 'five Ms') – money. During the course of much of my career, it was apparent that too many of us allowed ourselves to become dependent on others – accountants and quantity surveyors – for this vital part of our business. These specialists generally saw their part of the procurement process purely in terms of making or saving money on the profit and loss account. There was therefore a tendency to seek the cheapest solution at each stage of procurement, rather than the best overall value solution.

Institution framework

The Institution offers the following descriptors for this attribute in Appendix A of *ICE3001 Routes to Membership*:

5. Commercial Ability
Prepare and control budgets
Sound knowledge of statutory and commercial frameworks within own areas of responsibility

and additionally for Chartered status

High level of commercial and contractual understanding and an ability to use it

These few words are deceptively simple. In fact they cover a very wide range of understanding and ability, as evidenced by both the Development Objectives and the topic areas from MGNs 21 and 22. As a result, this attribute has been divided between three chapters of this book:

Chapter 9 covers the statutory framework in the UK (excluding Scotland)
Chapter 10 covers contractual framework.

This chapter covers the basics of commercial understanding. The relevant commercial Development Objectives are:

B2 Feasibility assessments and costing
- Value engineering
- Whole life costing

C2 Control budgets, tasks, people and resources
- Management systems, coordination
- Payment processes
- Project funding and financing

MGN 22 for all Members adds the topic areas

- rethinking construction
- infrastructure maintenance
- operational and maintenance cost analysis
- the financial implications of decisions made by engineers
- methods of funding of construction projects
- payment and compensation
- cost control
- estimating/tendering.

MGN 21 adds further aspects for potential Chartered Members

- whole life asset management
- methods of procuring construction projects
- budget management
- marketing
- private finance
- the financing of infrastructure development
- risk analysis
- supply chain management.

Sir John Egan DoE Report *Rethinking Construction*

A very good Executive Summary of this report is available on the official government website (see 'Useful websites' in the 'References'). In 1998, Egan, the former Director of Jaguar Cars, was asked to look in depth at the UK construction industry and offer advice on what could be done to improve its performance. His report followed *Constructing the Team* by Sir Michael Latham in 1994.

Egan identified five key drivers of change

- committed leadership

- customer focus
- integrated processes and teams
- quality driven
- commitment to people.

He identified a further four key elements where he felt performance could be improved

- product development
- project implementation
- partnering the supply chain
- production of components.

He realised that such fundamental reforms would require a culture change, including

- decent, safe working conditions
- improved management and supervisory skills
- design for ease of construction (standard components and processes)
- long-term relationships (not competitive tendering)
- clear measurement of performance
- sustained improvement in quality and efficiency.

To drive change, he set challenging targets for year-on-year improvements. He wanted KPIs to benchmark the industry for such things as cost, time, safety, waste and client satisfaction.

The report spawned a plethora of quangos (quasi autonomous non-governmental organisations) to drive the changes needed, including Movement for Innovation (M4I) and the Local Government Task Force. It was soon realised that this piecemeal approach would not work and, in 2003, they combined to become Constructing Excellence in the Built Environment (more often known simply as Constructing Excellence).

Eleven years later progress had undoubtedly been made, but few if any targets had been met. A survey by the *Rethinking Construction* group in 2009 (Wolstenholme *et al.*, 2009) found that, of 1771 firms surveyed, only 12% (representing 32% of the industry turnover) were using partnering, 10% were using KPIs and only 9% were accredited by Investors in People. As a broad generalisation, civil engineering firms have achieved more targets than many building firms, and the public sector and consultants have a higher awareness than contractors and manufacturers.

These initiatives continue to have an influence on the way projects are delivered; they therefore have a direct effect on your work. You should know how your organisation fits within these statistics and what it is doing to achieve best value. It is wise to familiarise yourself with the organisations and reports, which are freely available via their respective websites.

Traditionally, civil engineering organised its resources around projects. What innovation did happen was largely funded by individual project needs and there was a problem with capturing the innovation for the benefit of future projects. At last the industry is overcoming this serious weakness, making increasing use of the expertise of the whole supply chain to achieve best value.

In 1983, the Business Round Table in New York stated,

> All too often chances to cut scheduled time and costs are lost because construction operates a production process separated by a chasm from financial planning, scheduling and engineering or architectural design. Too many engineers, separated from field experience, are not up to date about how to build what they design, or how to design so structures and equipment can be erected most efficiently.

This was just as true in the UK. My job appointment in 1970 actually stated, 'We have plenty of people here who know how to design bridges, but not enough who know how to build them': the words of a County Bridgemaster (grand historical title) who was perhaps ahead of his time on what has become known as value engineering.

Value engineering

Best value is essentially focused on function: what does the client actually require of the solution and how can that best be achieved?

Value engineering is an organised methodology to identify and select the lowest life-cycle cost options which provide the required function consistent with the required performance, operation and maintainability of the project. It seeks to eliminate unnecessary costs in every aspect of the life of a project ('whole life costing'). This requires a joint effort from all of the many contributors to a project, from client and designer, suppliers and contractors, through users, to those responsible for maintenance, repair and replacement.

The client must clearly define acceptance parameters very early in the procurement programme, something which, in the past, many have been unable or unwilling to do. It also requires the client to accept that the capital cost may increase in order to reduce the overall life-cycle costs. A client may require the design life of a structure to be 100 years. What does that mean? How much maintenance and replacement is anticipated to achieve that life? What happens if more (or less) maintenance or replacement is necessary within that life span? Where will any extra resources come from? Certainly, insurance companies seem reluctant to underwrite consultants who are being asked by clients for guarantees of effective performance, so there is a significant risk.

To properly engineer value requires a joint effort by cross-functional teams (which has led to such arrangements as Early Contractor Involvement) and relevant suppliers (supply chain management). Inexperienced engineers should not merely partake in such arrangements, but think about and discuss why such cooperation was agreed and how effective it is. In this way experience is gained for future use, when they have to decide how to deliver increasingly better value ('more for less').

The continuing quest for efficiency savings and improvements to the whole procurement process requires engineers to make judgements based on a complex and wide range of criteria, far beyond the financial considerations. Feasibility studies and cost/benefit analyses are certainly an important part of the search for greater efficiency and better quality, but money is not the sole arbiter. Best value is not the same as lowest cost and, in a society which is realising that it may have less to spend, clients and providers generally recognise this fact. But they all do want better value for investment, delivered within budget and on time, as well as a predictable lifetime costing. To do this efficiently and economically, every asset must have a management plan for its entire life.

Whole life asset management

Whole life asset management is a more rational system for deciding on the best overall value of an asset. Compare this with the system of doing (as economically as possible) each part of the survey, feasibility, design, procurement, operation, maintenance, modification and rebuild as a separate operation. Economies at each stage can have significant knock-on effects, which can cause delayed, but greater, expense. By integrating the entire process, the most effective methods can be chosen at each stage to minimise the total cost, whether or not they are the most economic for each particular stage.

The disadvantage of whole life costing is that it may often require more capital investment upfront (early borrowing with associated interest charges). This can make it more attractive at times of lower interest rates.

Many organisations have systems for the routine management of assets. If you become involved in any way, then you must familiarise yourself with those systems. To assist in understanding the decision-making process, there follow certain rudimentary steps common to any asset management plan, at each of which judgements must be made to provide meaningful answers.

(*a*) What did the asset cost? This figure should include the interest on any loans used in the purchase.
  - How much would it cost to replace? Where would the resources to replace it come from?

  - What would be its residual value if it had to be sold? Is there a market for it (if there is not, then has it any value)? See also the section on 'Balance sheet' under 'Company financial statements' below.
  - Has the depreciation in value of the asset been used to provide a sinking fund for its repair and/or replacement?

(*b*) What use is the asset?
  - Is the purpose still the same as when the asset was acquired?
  - Is the asset still effective in fulfilling that need?
  - Could the asset be made more effective by modification?
  - Could the current purpose now be achieved in a different way altogether?

(*c*) How old is the asset?
  - What is its condition?
  - How much longer will it last?
  - What maintenance and/or refurbishment does it need to keep it operational?
  - What maintenance or upgrading does it need to keep it safe in use and compliant with current legislation and standards?

(*d*) What are the associated costs?
  - Consumables, routine maintenance, modifications to ensure compliance with changing laws and standards and increasing user expectations.

  When these costs are properly brought into an asset management plan, they do tend to show that a more costly asset provides better long-term value (i.e. higher initial capital expenditure can reduce routine revenue expenditure).
  - Could the capital invested in ownership of the asset have been used more effectively elsewhere to produce a better return?
  - Could the asset have been leased or sub-contracted to retain more flexibility and reduce the capital expended?

  These last two points require a balance between the potentially higher cost over a longer period of lease (hire purchase), the high initial outlay of ownership and the alternative use to which the money could have been put. A politically attractive option, during a period of particularly low interest rates, the Government has embraced Private Finance Initiatives (PFIs) or Public Private Partnerships (PPPs) to give them greater short-term financial flexibility but, arguably, less control of workload, performance and quality.

In a similar mode, your employer may well have contracted out the provision of many of its services. Refreshments, company vehicles, office cleaning, and information and computer technology (ICT) are frequently outsourced.

Budget management and cost control

Because of the overriding importance of financial management, all organisations have processes and procedures for acquiring money and controlling expenditure. It is most important that every potential professional engineer gets to know them and understands how to use them. You will inevitably be involved in fundamental aspects of budget control, even if it is something as simple as completing timesheets, so find out how these are used to control expenditure.

Review candidates often say in their reports, 'I monitored expenditure against the budget', which suggests a purely administrative role; it tells the Reviewers nothing about their commercial responsibility. What is needed is some explanation of how the candidate judged when the degree of divergence became unacceptable and what they did about it.

Candidates for membership should know the sources of the money for their projects, and how the necessary resources were procured, even if they had no part in obtaining them. The financiers of a project must be told routinely how their contribution is being spent; there will be systems for doing this, so find out about them. I remember a multi-million pound project with eight different sources of funding, each of whom had to be kept informed of the efficient expenditure of their contribution.

Know what your own costs are, and what percentage of your salary constitutes your charge-out rate. What do these rates include (e.g. ICT and administrative support, as well as office space and equipment, insurance, sickness and holidays)?

(An important point here for Review candidates: never divulge actual figures, which are, of course, commercially sensitive. Always give them as percentages of your salary, which the Reviewers are unlikely to know. However, they *will* be aware of the approximate percentages, so make sure you are correct!)

If, for example, you commission a soil investigation, a survey or some temporary works (from either internal or external specialists) how was the work paid for? And how did you decide what it was reasonable to spend (the budget)? External specialists are usually employed under contract, so it is relatively easy to find out. But how are internal specialists reimbursed?

What is your organisation's annual profit on turnover? Many organisations repeat the mantra that 'our people are our greatest asset', but it is also probable that staff account for the majority of turnover and are 'our greatest cost'. These figures are in the public domain for the vast majority of organisations, so read and comprehend their annual statements (see below), very often placed in the company foyer for visitors to read. How, for example, does the profitability of your part of the business compare with the organisation's average? Can you account for any divergence?

Company financial statements

Finance and its implications underpin all key engineering decisions, but battling your way through a set of company accounts can be challenging, if not baffling, unless you understand their basic components and language. To understand financial documents you need a basic knowledge of the terminology.

All European Union listed companies use the International Financial Reporting Standards (IFRS). Companies not so listed may use UK Generally Accepted Accounting Principles (UK GAAP) which, although similar, include detailed differences in the manner in which some items are recognised, measured and presented. All UK-based companies must comply with these standards as appropriate. Many provide additional information beyond these standards.

Notes to the accounts

These provide background information, such as how the value of an asset has been derived. Perhaps surprisingly, they are often at the end of the financial statements, yet they are a vital precursor to understanding the accounts. Reading them before looking at the detail enables you to understand the judgements underpinning the accounts.

Main types of company accounts

There are three main accounts

- profit and loss (P&L)
- balance sheet
- cash flow statement.

Profit and loss account

Sometimes known as the income statement, this account outlines how the money received from the sale of services or products is transformed into income, disclosing how much has been earned during a year and what is available to invest or give back to shareholders.

Revenue (known also as 'turnover') is the amount the organisation actually makes through the sale of its services or products. Companies sometimes also provide information on underlying revenue, excluding revenue from acquired businesses and the disposal of assets, which is a useful indication of how the core business is doing.

Cost of sales is directly attributed to the production of whatever the company delivers or produces. It includes the salary costs of all employees directly involved.

Gross profit is the difference between revenue and cost of sales. This margin that the company makes is a good indicator of the viability of the organisation. If the company has made a loss, the figures are shown in brackets.

Gross margin shows the gross profit as a percentage of revenue and is a useful way of comparing the financial performance of different organisations operating in the same field of work. However, this figure should be treated with caution, because the margin may decrease while both revenue and profit increase, perhaps as the result of the acquisition of a company achieving lower margins.

Operating profit is the net income earned from core business operations, after deduction of both direct costs of sales and indirect expenses, which include things such as support functions and pension costs. Some companies include earnings before interest, tax, depreciation and amortisation (EBITDA). This measure of profit (non-GAAP) is essentially operating profit with depreciation and amortisation (spreading the cost/value of assets over their useful lives) added back in. But do read any notes to the accounts carefully; companies tend to interpret these measures differently.

Profit for the year takes account of any income or expenses relating to the way the business is financed (e.g. the interest accruing on loans and income tax). So this is the profit left, which could be paid out to the owners of the company (as dividends to the shareholders or partners), some of which may be retained by the company for future acquisitions.

Balance sheet

This account is a statement of what the business owns (assets), less any amounts outstanding payable to other parties (liabilities), on a particular day, usually either 31 December or 31 March.

Assets are categorised into current (those which can be converted into cash in less than a year – liquidity) and non-current. 'Working capital' is current assets less current liabilities. Cash and cash equivalents (e.g. short-term government bonds and Treasury bills) are the most liquid assets, since they already are, or within three months can be converted into, cash. Other assets were traditionally valued at the price for which they were acquired but, recently, in more difficult trading conditions, the concept of 'fair value' has been used. If the value of an asset has to be reduced to better reflect its current market value, it is known as an impairment or write-down.

Intangible assets include such things as lists of clients, key personal contacts, patents and brand names. Their costs are spread over the period in which they are considered to have value (amortisation – see earlier).

Property, plant and equipment (PP&E, not to be confused with PPE – Personal Protective Equipment) are physical assets that are used in day-to-day activities, and include buildings, machinery and IT equipment, all of which depreciate with time, so provision is made to spread the cost of the asset over its useful life and provide for replacement.

Inventory is stock held by the business to be either sold on or used in manufacturing, construction or development. It is generally held for as short a time as is reasonable, since it ties up capital. Many companies aim for 'just in time' delivery of materials and components to minimise their inventory.

Liabilities are obligations that a company must settle, normally in cash, and are divided into current (e.g. invoices) and non-current items.

Provisions are liabilities for which either the timing or amount is uncertain, such as litigation or restructuring costs. At the time of writing, there are significant increases in provisions in many companies because of their pension fund deficits.

Loans and borrowings are liabilities for which the interest appears as an expense on the P&L account.

Equity is the remaining value of assets when all liabilities have been accounted for, and is the amount to which shareholders have claim. It is made up from the initial value of the shares issued plus the earnings that have been retained by the company over the life of the organisation. It does not include any earnings paid out to the shareholders as dividends.

Cash flow statement

In contrast with the P&L account, where income is recorded when it is earned and outgoings when the liability arises (whether or not it has been received or paid), net cash flow is the *actual* cash received less the cash paid out.

The cash flow statement indicates where cash is coming in and how it is being used in running the company. It determines the organisation's ability to meet its short-term liabilities, such as the payment of outstanding invoices or the repayment of a loan. Companies that appear on paper to be profitable and to have positive net assets can fail because they cannot pay their immediate bills. Tough trading conditions may cause a company to go into administration simply because the banks have called in loans and there was insufficient cash to cover the repayment.

The majority of companies choose to spend or invest most of their cash flow to enable them to make income in the future, by acquisitions, purchasing new plant and equipment or repaying loans, and to retain capital and good faith by paying better dividends to their shareholders. But, to balance these desires, they must retain sufficient cash or cash equivalents to cover possible contingencies.

Estimating and tendering

Any organisation must win work in a competitive market at a price that enables it to stay in business by making a profit. Even in the public sector, estimates must be made for inclusion in the routine (annual) budget review and, if successful, the projects then have to be delivered within that budget estimate.

So, every engineer must understand how to realistically quantify the work that will be required.

- Decide what resources are needed, when and for how long.
- Check the ideal resource allocation against what is likely to be available. This may mean that resources have to be recruited or engaged from beyond the organisation itself.
- Decide how the resources will be financed – sources, costs, overheads, expenses, loan interest, etc.
- Anticipate possible problems (contingencies) and allow for the financial risks involved. This requires assessment of their possible effect on progress, resources and cash flow and how likely they are to happen.
- Relate running costs to the frequency of payment, allowing for any retentions (cash flow).

And, after all these judgements, the company must still achieve a bid which stands a chance of success! A sound knowledge of the current marketplace is vital – without running any risk of being suspected of collusion.

These judgements are just as relevant in surveys, investigations, feasibility studies, design, maintenance, demolition, research, lecturing, self-employed consultancy, etc. as in construction, so all civil engineers, whatever their business, must know about them.

Initial Professional Development for Civil Engineers
ISBN 978-0-7277-4147-9

doi: 10.1680/ipdce.41479.059

Chapter 9
Statutory framework

The statutory framework is the law established by the ruling authority in a country, which is made by Parliament in the UK, and enacted by the monarch's signature (the UK is a constitutional monarchy).

Statutes cannot be ignored anywhere within the jurisdiction of that ruling authority. A simple example can be seen at the till of UK shops where there is often a notice explaining that the store's guarantee does not affect your statutory rights. The store's policy on refunds and replacements cannot be less advantageous than the rights set out in law (in this case, by the Sale of Goods Act 1979).

The law covers an enormous range of situations through a wide variety of courts and methods. To understand the legal systems to which civil engineers and civil engineering firms must conform in the UK, it is necessary to have a broad overview of the different categories of law. A similar view is necessary if you are working in jurisdictions beyond England and Wales.

Institution framework

The relevant part of the Commercial attribute in Appendix A of *ICE3001 Routes to Membership* is:

Sound knowledge of statutory . . . frameworks in own areas of responsibility

What is 'the law'?

The law is a formal mechanism of social control in a society; formal because it can be enforced through the legal system and the courts (i.e. it is a mandatory requirement). It applies throughout a society (usually a country) to the inhabitants generally.

Other rules apply only to groups in limited situations, such as games and contests, religious beliefs and many professions, but in those cases there is no legal sanction to force compliance or to punish non-compliance, beyond the disapproval of that community (i.e. it is an obligation).

The Institution of Civil Engineers has its Code of Conduct, which all potential Members signs to affirm that they will comply with its terms when elected, but it is not a law. Thus, it is obligatory under Rule 5 that we all record our Continuing Professional Development, but it is not mandatory (i.e. there are no penalties in law). Sanctions for non-compliance with the obligations of the Code range from being asked by the Institution to bring your records up-to-date within a predetermined limited period, having your name and misdemeanour published in an appropriate journal (such as *NCE*), to being suspended for a period or even, in the worst cases, rejected by the Institution, thus being unable to use their designatory letters.

Consider first the distinction between international and national law.

International law

Much of international law comes from treaties agreed between the governments or rulers of countries. If you are working in a foreign country, it is important that you both understand and obey the laws of that country and have an adequate overview of any relevant treaties between that country and your homeland (for example, reciprocal medical agreements or the laws on alcohol and public behaviour).

The European Union fits somewhere between national and international law because, although it is an entity, it consists of nation states. There is more about its increasing influence on English law later in this chapter.

National law

There are wide differences in law between individual countries. As an example, Scotland has its own law and legal system, which is quite distinct from that of England and Wales. The jury in a criminal trial, for instance, has 15 members in Scotland and can reach a verdict by a simple majority of 8 to 7, while in England and Wales there are 12 jurors, at least 10 of whom must agree a verdict. This book does not cover Scottish law.

Within UK national law there is a distinction between public and civil law. Public law involves the state or government in some way (which is why 'Regina' appears so frequently in case histories), while civil law is concerned with disputes between individuals or businesses.

Public law

Public law falls into three main types:

Constitutional law – which controls the method of government and any disputes that arise over, for example, who can vote in an election or become a Member of Parliament, or whether the correct procedures were followed at an election.
Administrative law – which controls how ministers of state or other public bodies

such as local authorities and councils should operate. An important part is the right to judicial review.

Criminal law – that sets out the behaviours that are forbidden at risk of punishment. A person who commits a crime is said to have offended against the state, and so the state ('Regina') has the right to prosecute them, even where the victim can bring a private prosecution against the perpetrator. However, as well as punishing the offender, the courts have the power to order the criminal to pay damages to any victims.

Civil law

Of the many different branches, the main ones of concern to civil engineers are detailed below.

Law of contract – covers legal relationships voluntarily entered into, such as

- performance which does not match legitimate expectations
- goods which are found to be not fit for purpose
- failure to pay predetermined instalments, such as interim valuations or hire purchase instalments.

Contract law is covered in greater detail later in this chapter.

Tort (literally 'twisted, crooked' from Latin) – occurs where the law holds that one person has a legal responsibility to another person, even though there is no contract between them. A tort is a civil wrong, other than a breach of trust or breach of contract, and covers, for example

- trespass – intentional and direct interference with another's person, property or land
- nuisance – indirect interference with another's land
- negligence – unintentional and careless interference with another's person or property
- defamation – slighting of another's reputation.

Some examples will help to clarify the above definitions

- a passenger is injured in a vehicular collision (the tort of negligence)
- a household complains that it is being adversely affected by the noise and dust from a construction site (the tort of nuisance)
- an engineer freely criticises an architect without just cause in fact (the tort of defamation)
- a person is injured by faulty machinery on site (the tort of negligence, but also may involve occupier's liability and/or employer's duty under the Health and Safety Regulations).

Table 1. Differences between criminal and civil law

	Criminal cases	Civil cases
Venue	Magistrates' court or Crown Court	High Court or county court Some cases, notably family, may be heard in magistrates' court
Decision by	Magistrate or jury	Judge (or panel of judges) Very rarely a jury
Person bringing charges	'Prosecutor' Crown Prosecution Service or other state agency, such as Environment Agency	'Claimant' (formerly 'plaintiff') – the individual whose rights have been affected
Result	'Guilty' or 'not guilty' or 'convicted' or 'acquitted' The criminal is punished: ■ prison ■ fine ■ probation ■ community service order ■ curfew order, etc.	'Liable' or 'not liable' to put the matter right as far as possible by: ■ compensation (damages) ■ injunction, to prevent similar actions in future or ■ an order for specific performance where the defendant broke a contract and is ordered to complete the contract
Proof	'Beyond reasonable doubt'	'On the balance of probabilities' – a lesser standard, so even though a person has been acquitted in a criminal court, a civil case, based on the same facts, may subsequently be successful

Other branches of civil law which commonly involve our industry include:

company law – which regulates how a company must be formed and run
employment law – which covers all aspects of employment, from contracts of employment to redundancy and unfair dismissal.

There are other aspects of civil law, too numerous to mention, which you are unlikely to come across in civil engineering. The major differences between criminal and civil law are detailed in Table 1.

The evolution of English law

Historically, the most important methods of keeping the peace were established custom, initially at tribal, local or regional level, and later the decisions of judges, which started to create a national uniformity (or 'common law').

As Parliament became more powerful in the eighteenth and nineteenth centuries, Acts of Parliament (statutory law) were the main source of new law, but the decisions of judges (case law) were still important in interpreting the Parliamentary law and filling gaps where no statute law existed.

During the twentieth century, statute law and case law continued to be the major sources of law but, in addition, two new sources became increasingly important – delegated legislation and European law.

Delegated legislation

This is law made by some person or body other than Parliament, but with the authority of Parliament, usually laid down in an 'enabling' Act, which creates the framework of legislation. The power is delegated to others to make more detailed law within the framework.

This power takes a number of forms:

Orders in Council – can be made by the Queen and Privy Council under the Emergency Powers Act 1920, usually only enacted in times of national emergency

Statutory Instruments – give authority to ministers and government departments to make regulations for areas within their particular sphere of responsibility. For example, the Health and Safety at Work etc. Act 1974 now has a large number of *Regulations under the Act* made by various Ministers of State, such as the Construction (Design and Management) Regulations. European Directives are mainly implemented in the UK as Regulations

Byelaws – can be made by public corporations and certain companies for matters within their jurisdiction which involve the public, such as civil enforcement of parking restrictions.

European law

The UK joined the original six nations of Europe in January 1973, in what was then the European Economic Community. It became the European Union (EU) in 1993, and currently has 27 Member States.

The aim of the EU is to achieve greater cooperation and cohesion in trade, economics and standard of living through harmonisation of the laws of the Member States. It has had a particular impact in the UK on employment and equality, but also, for example, now requires construction contracts above certain values to be open to tender by firms throughout the union.

EU law takes precedence over UK laws; not simply any laws enacted since the UK joined the EU, but also those laws in place before that date.

Europe has two sources of law – primary and secondary.

Primary sources of law

Principally, these comprise the Treaties signed by all national governments, which are binding and become part of English law automatically. This means that UK citizens can rely on the rights inherent in a Treaty, even though those rights may not have been specifically enacted in English law, and can challenge any contravention in an English court.

Secondary sources of law

Legislation passed by the Council of Ministers, in the form of either Regulations, Directives or Decisions, upheld by the European Commission and the European Court of Justice.

> *Regulations* are 'binding in every respect and directly applicable in each Member State' (Article 249 of the Treaty of Rome). For example, when the UK government was reluctant to legislate for tachographs in lorries, preferring a voluntary agreement with lorry owners, the European Court of Justice ruled that Member States had no discretion.
> *Directives* 'bind a Member State to the result to be achieved, while leaving to domestic agencies a competence as to form and means'. Thus, the UK passes its own laws to implement Directives, usually within a time limit imposed by the European Commission. The usual UK method is by Statutory Instrument. For example, the Working Time Directive, detailing maximum hours to be worked, rest periods and paid holiday, which should have been implemented by November 1996, passed into UK law two years late in the Working Time Regulations 1998. Not all Directives are implemented in this way: for example, a Directive on liability for defective products, issued in July 1985 for implementation by 30 July 1988, was implemented by Parliament in the Consumer Protection Act 1987.
> *Decisions* may be addressed either to a Member State or to an individual company or person, and are 'binding in every respect for the addressees named therein'. They are generally administrative in nature.

The court structure

Most civil cases in England and Wales are not heard in the civil courts, but in alternative forums such as conciliation and arbitration, all of which are covered in Chapter 10.

Criminal offences fall into three categories

- *summary offences* – minor offences tried only in a magistrates' court
- *indictable offences* – serious offences tried only in a Crown Court
- *either-way offences* – intermediate offences that can be tried summarily in a magistrates' court or, on indictment, in the Crown Court.

Magistrates' courts

The majority of cases are tried in the magistrates' court. Magistrates' courts have restricted civil jurisdiction over minor family matters, minor debt (e.g. non-payment of Council Tax) and the granting of licences (such as to taxis, public houses and nightclubs).

Almost all criminal cases commence in a magistrates' court and over 95% are resolved there. Magistrates have limited sentencing powers; where they believe an offence merits a more severe sentence, they commit the offender for sentencing at the Crown Court.

Outside London, magistrates are lay justices (Justices of the Peace (JP)), sitting in twos or threes, advised by a justice's or court clerk. There are over 30,000 magistrates, who sit part-time on around 350 benches in England and Wales, and continue a long tradition of amateurs ensuring that the common sense and values of ordinary people are reflected in the justice system.

The Crown Court

There is only *one* Crown Court, divided between some 90 centres throughout England and Wales. It is the criminal court which deals with the most serious (indictable and some either-way) offences (less than 5% of the total).

Cases are heard by a High Court judge, circuit judge or recorder, depending on the gravity of the case. The Court also hears appeals from summary conviction in the magistrates' courts and sometimes matters such as licensing appeals from civil jurisdiction.

County courts

There are around 260 county courts, which are cheaper alternatives to the High Court, dealing exclusively with civil actions. Their case-load consists of contract, tort (especially personal injuries), property, divorce and other family matters, bankruptcy, equity and race relations, etc. Many cases are claims for overdue debts.

One important distinction between the county court and other courts is the right of solicitors to be heard in the county courts.

Small claims in the county court (under £5000) are processed through a special, simpler (less expensive) procedure. Legal representation is not required, and the district judges adopt an interventionist role, dispensing with formal rules of evidence and being more inquisitorial than normal.

The High Court

This is situated in the Royal Courts of Justice in the Strand, London and within 24 of the provincial Crown Court centres. It has four divisions:

- *The Queen's Bench Division* is the largest, generalist division, consisting of the Lord Chief Justice and 70 lower-ranking judges, dealing with common law such as tort, contract, debt and personal injuries, as well as two specialisms – admiralty and commercial.
- *The Chancery Division* consists of the Vice-Chancellor and 18 lower-ranking judges, dealing with claims relating to property, trusts, wills, partnerships, intellectual property, taxation, probate and bankruptcies. It has two specialist courts, Patent and Companies, and sits in eight of the provincial Crown Court centres and London.
- *Divisional Court* hears appeals and exercises supervisory jurisdiction, reviewing the legality of both inferior courts and the executive.
- *The Family Division* consists of a President and 17 judges, who hear divorce cases and ancillary matters and cases under the Children Act. It has over 50 provincial centres.

There is a number of specialist courts, dealing with complex cases, such as restrictive practice, fair trade and the property of those lacking mental capacity. One of some importance to construction is the Technology and Construction Court, which takes complex technical or factual cases (mainly building contracts and computer-related disputes) from the Queen's Bench or Chancery Division to be heard by circuit judges.

Initial Professional Development for Civil Engineers
ISBN 978-0-7277-4147-9

doi: 10.1680/ipdce.41479.067

Chapter 10
Contractual framework

Every candidate for the Professional Reviews is expected to understand the broad basis and key elements of the contracts they will work with, such as how the risks are apportioned and the systems for ensuring that the contractor is recompensed. I would include your own contract of employment in this. You will not be asked any detailed questions which explore areas well beyond your direct experience. So, if you have never worked on a Joint Venture, you will not be asked about these in any detail.

This book cannot discuss the many standard and non-standard forms of contract in use in the industry, only the basic principles applicable to all of them.

Institution framework

The Institution offers the following descriptors for this attribute in Appendix A of *ICE3001 Routes to Membership*:

> **5. Commercial Ability**
> For Chartered status
> High level of . . . **contractual understanding** and ability to use it

MGNs 21 and 22 add further to the topic areas:

> Partnering/alliances
> Site/project management
> Performance specifications
> Forms of Contract for civil engineering works
> Joint Venture contracts
> Target cost contracts
>
> Rethinking Construction
> Marketing
> Supply chain management

Law of contract

A contract is any agreement which those people making it (known as 'the Parties to the Contract') intend to be binding on them all and does not need to be signed or even written to be binding upon its parties. The legality of a contract depends only on whether they intended to make a bargain and what that bargain was intended to be.

The rules to make a contract enforceable in law (a 'valid contract'), termed 'essential principles', have been established mainly by common law (refer to Chapter 9), based on existing practice and precedents (case law). Statutory law does restrict what agreements can be made (for example, it is illegal to enter into a contract to do something which is, in itself, illegal).

The essential principles are suggested by *Chitty on Contracts* (Joseph Chitty was one of a dynasty of lawyers active from 1775 to 1899 – his book is now in its thirtieth edition and encompasses over 180 years of case law).

- Privity of contract – a contract cannot apply to anyone not party to it.
- Objectivity – the parties must make choices on merit, not prejudice or favouritism.
- Contractual interpretation – the wording must be complete and precise, as far as possible, and not open to ambiguity or differing interpretations.
- Freedom of contract – the parties cannot be coerced into entering the contract.
- Binding force – once the parties have signed up to a contract (which may be merely shaking hands) then the law expects each of them to expeditiously complete their part of the bargain.

Verbal contracts

It is obvious that, in any verbal contract, it is much more difficult to prove what was actually agreed, so these are best avoided. You may believe that all your organisation's work is under written contract, but there are two situations which might inadvertently cause problems.

(*a*) Letters of intent – where it is not deemed possible to execute a contract straightaway, a letter of intent may be issued. These can be the source of enormous commercial risks. If your organisation allows work to start under a letter of intent, then you should find out how the risks have been defrayed.

(*b*) Acceptance inferred by conduct – it is tempting to start work, based on cordial relationships, in anticipation of the contract documents, or a contractor or client may send an order requesting that you start work within a short time. Avoid starting work before replying, as it is then too late to query anything in the order. However, if a formal contract is subsequently signed, then the terms agreed apply retrospectively to all work done under that contract.

Written contracts

Written contracts may simply be signed by the parties or formally executed under seal. The principles are largely the same, with one or two specific exceptions which are beyond the scope of this book.

The courts will enforce a contract (whether written or not) if the following four steps can be shown to have occurred

- offer
- acceptance
- consideration
- intent.

Offer

An offer must be clearly communicated and capable of acceptance. A client's invitation to tender, like a display in a shop or forecourt, is an invitation to make an offer ('invitation to treat' in legal jargon), not an actual offer.

A tender is an offer if it contains all of the details (such as price, scope and personnel) necessary for clarity and capability (one of Chitty's principles). If this is so, the client can form a contract simply by picking up the phone and saying that the tender is accepted.

Acceptance

Acceptance of an offer must be communicated clearly and unequivocally. Any request by the client to vary the tender (e.g. changes or a request for a reduction in price) creates a counter-offer for consideration by the tenderer. Subsequent changes suggested by either party create further counter-offers. It is possible to tacitly accept a counter-offer by commencing the work (see above), so care is needed at this stage.

A signed contract is the best evidence of exactly when acceptance occurred and the contract was formed.

Consideration

For a contract to be binding, 'consideration' or benefit must be present. For example, a consultancy appointment requires the consideration from the consultant to be the deliverables (e.g. calculations, requisite approvals, drawings, specification, contract documents, supervision, etc.) and from the client the payment of fees. In law, the consideration must have some value but need not be sufficient (for instance, a peppercorn rent, which bears no resemblance to the true costs). This means that a contract is still enforceable even if it is a really bad deal for one of the parties.

The 'consideration' must not have been used before – 'past consideration is no consideration', so variations attract further payment. To expect more without fresh consideration (an increase in payment and/or time) is not enforceable in law.

Intention to create legal relations

The courts assume that the parties to a business contract intend it to be legally enforceable. If, however, the parties do not want to be bound by the arrangements (e.g. a memorandum of understanding), the intention not to be legally bound must be stated explicitly.

Statute law in construction

Not all the thousands of extant Acts of Parliament directly affect the construction industry, but some which do include

- Health and Safety at Work etc. Act, first enacted in 1974, from which derive many Regulations (see Chapter 9 under Delegated legislation), including the Construction (Design and Management) Regulations
- Housing Grants, Construction and Regeneration Act
- Contracts (Rights of Third Parties) Act
- Arbitration Act
- New Roads and Street Works Act
- Occupiers' Liability Act
- Supply of Goods and Services Act.

As you gain experience, you should make notes of all Acts and Regulations which influence your work, including current dates. There is one in particular that needs to be discussed here – the Housing Grants, Construction and Regeneration Act, introduced in 1996. At the time of writing (2010), all political parties are in agreement that a new Act is required, but my information is that the rules will remain substantially the same.

Housing Grants, Construction and Regeneration Act 1996

Michael Latham's report *Constructing the Team* (1994) stated that a better way of resolving disputes than through recourse to the law had to be found in view of the adversarial nature of many construction contracts. The government, in the interests of legislative speed, tacked a section onto the Housing Grants, Construction and Regeneration Act 1996 (HGCR Act), which had two major effects on construction

(*a*) payment procedures were defined
(*b*) automatic right of adjudication was introduced.

Payment procedures

The Act requires fixed dates for payments, with notice of withholding sums due and rights to suspend performance due to non-payment. It effectively outlawed 'pay when paid' clauses – the cause of much discontent until then.

Right to adjudication

The Act sets out requirements which enable a party to refer a dispute at any time to an adjudicator, who must then act impartially to ascertain the facts. The courts will not consider a case unless adjudication has failed to reach agreement. Adjudication is covered in more detail below.

There are other ways of settling contractual disputes which are less costly and time-consuming than adjudication. The best of all is by avoiding disputes, by partnering to achieve a common purpose – a sound job at a fair price in a realistic time – something which my mentor was doing in 1958! The ICE firmly believes that its suite of Engineering Construction Contracts (ECC) is the means of achieving this, but any contract's success is reliant on the goodwill of all parties.

Alternative dispute resolution

The alternative dispute resolution (ADR) methods are considered below in increasing order of complexity and, therefore, cost.

Negotiation

Negotiation is still the quickest and cheapest method of settling disputes, with no one beyond the parties being aware of the details. It does need both parties to abandon entrenched attitudes, be prepared to give and take and be determined to reach an equitable agreement. This often requires someone very senior in both parties to insist that a settlement must be found.

If the parties cannot reach agreement, they can instruct solicitors to try to negotiate a settlement. Obviously, costs then escalate. Even when court proceedings have been commenced, the lawyers for the parties will often continue to negotiate, attempting to reach an out-of-court settlement.

Many cases are settled at the last minute, 'on the steps of the court', which does suggest that the parties only see sense when they

- face the high costs of a court case
- realise that the court is going to take a purely legal view, with no commercial or technical compromise, which is probably not in either party's best interests
- realise that they will generate adverse publicity.

Mediation

The disputing parties agree a suitable neutral person as mediator, whose role is to consult with each party to find as much common ground as possible, by exploring their positions, looking at their needs and carrying offers to and fro, while keeping confidentiality. Normally, mediators are facilitators and do not divulge their own views. But they can be asked for an opinion, in which case the mediation becomes more of an evaluation, but the primary aim is to end the dispute. Among a growing number of commercial mediation services, the main one is the Centre for Dispute Resolution in London, which claims that over 80% of its cases are successfully resolved.

Conciliation

The parties agree to appoint someone as a conciliator, who is expected to suggest grounds for compromise and a possible basis for settlement. Most of us in the UK have heard of the conciliation service ACAS (Advisory, Conciliation and Arbitration Service), a government-sponsored organisation which is often at the interface of union/employer disputes.

The conciliator holds confidential meetings with the parties separately in an attempt to resolve their differences, by lowering tensions, improving communications, interpreting issues, providing technical assistance and exploring potential solutions in an attempt to negotiate a settlement. The parties seldom actually meet in the presence of the conciliator.

The main goal is to remove animosity and distrust by seeking concessions. It differs from arbitration in that the process has no legal standing, the conciliator usually cannot seek evidence or call witnesses, nor do they usually write a decision or make an award. The conciliator tries to get the parties to agree among themselves.

Under the ICE's Conciliation Procedure, the conciliator's recommendation is deemed to have been accepted unless, within one month of its receipt, a written Notice of Adjudication or Notice to Refer to Arbitration is served on the other party.

Adjudication

The HGCR Act states that 'all parties to a construction contract have the right to adjudication at any time' and that a 'contract must contain a procedure which complies with the Act'. So, either party can opt for adjudication at any time. The courts will not consider a case unless adjudication has failed to reach an agreement acceptable to both parties.

The Act further states that the adjudicator should be a 'construction professional'. Many of the construction Institutions offer adjudicators, but currently only the ICE has a publicly available list of selected, trained and monitored adjudicators and a published procedure.

In the increasing complexity from negotiation to litigation, adjudication is the first stage in which the parties to the dispute relinquish control. The adjudicator has the power to call for evidence, to ask for further particulars and to control the process entirely.

Under the *ICE Conditions of Contract* 7th edition, the adjudication 'shall be conducted under the ICE's Adjudication Procedure (1997)', which requires a swift outcome (normally 28 days). The decision is final and binding, unless a party gives notice to refer to arbitration or the courts within three months from the decision. Other contracts, including the JCT form, have similar provisions.

There is a political intent to replace the provisions for adjudication in the HGCR Act with more appropriate legislation. However, the provisions are likely to be much the same.

Arbitration

Arbitration is the voluntary submission of a dispute between the parties to some person other than a judge. The agreement to go to arbitration can be made at any time – when the dispute becomes apparent or, more usually in construction contracts, before a dispute arises, as part of the contract. Where an arbitration agreement forms part of the contract, the courts will normally refuse to deal with any dispute – it *must* go to arbitration. The parties cannot withdraw from the process.

The parties are free to agree on the number of arbitrators and on the appointment procedure, but most contracts either name an arbitrator or provide a method of choosing one. The choice is frequently devolved to the President of the ICE or the Institute of Arbitrators, who usually chooses someone who has expertise in the particular field of dispute. In the last resort, the court can be asked to appoint an arbitrator.

There are many forms of hearing, from the arbitrator making the decision on the submitted documents alone, through to submitting all the documentary evidence before both parties attending a hearing, making oral submissions and calling witnesses to support a case. Most arbitrators prefer that evidence is given by those most involved and knowledgeable, not by legal representation. The witnesses are not normally on oath, unless the parties wish to make the hearing even more formal.

The decision, called an award, is binding on both parties and can, if necessary, be enforced through the courts. It can only be challenged on the grounds of serious irregularity in the proceedings or on a point of law. It cannot be challenged simply because one party or other disagrees with the decision.

There seems little doubt that businesses are moving away from legal resolution of disputes towards the alternatives, of which the best remains negotiation. After a residential course on construction law, a director of the company which sponsored me some years ago justified the cost by stating, 'If all it has done is make you determined to avoid legal disputes, then it has been well worth it'. Once the law gets involved, costs escalate disproportionately and any decisions are based purely on the legal arguments, with no room for compromise or common sense.

Tribunals

Tribunals are an increasingly important part of the legal system, relieving the courts of much casework, mainly relating to the welfare state. Out of more than 60 different types of tribunal, the most likely to affect construction is the employment tribunal.

Law of tort

The primary function of the law of tort is to define the circumstances in which a person whose interests are harmed by another may seek compensation. Therefore, many clauses in a contract are designed to incorporate the likely obligations and remedies of tort.

The elements of negligence are

- Duty of care is owed. To whom does your employer owe a duty of care?
- Standard of care is not reached and damage has been suffered.
- Reasonable care must be taken to avoid acts or omissions which can reasonably be foreseen and which would be likely to injure persons who are closely and directly affected by the work.
  - It is not enough to put a sign up at the boundary of a construction site saying 'Danger Keep Out'. There must also be a 'reasonable' fence to prevent entry.
  - Dust suppression must be seen to be 'reasonable'. It is not enough to spray a temporary access road with water once, first thing in the morning, if the day is hot and the road is in constant use.

The law rarely expects a professional always to achieve the desired result, but expects that reasonable skill and care will be used. Surgeons do not *have* to cure a patient and solicitors cannot win every case, but we are all required to use our best professional efforts (see also Chapter 14).

The law and application of contracts are complex and vitally important to every organisation. Most therefore have experts (which might be legal advisers or the Company Secretary) from whom advice and guidance are readily available. Relatively inexperienced engineers will be reliant on their expertise. But remember that they are *advisers*. No doubt they take their responsibilities to supply correct advice very seriously (as do health and safety advisers), but the responsibility for compliance and successful application remains yours, as the organisation's engineering representative.

The following list defines areas for which every organisation has procedures and processes, and it is not exhaustive. Make sure that you know and understand what your company's procedures are and how they are applied, particularly if they have direct relevance to your involvement. You may not be responsible just yet, but inevitably you will be as you climb into more managerial roles.

- For clients that are part of a corporate group, exactly which group company is signing the contract? What happens if, as is not uncommon today, the company is bought out?
- Your organisation will have systems in place to identify and avoid clients or contractors in such a poor financial state that there is a significant risk of them not paying invoices. When a company becomes insolvent and is wound up, payment of invoices is low in the hierarchy of creditors, so can be a significant financial loss.
- Only companies that have the financial strength to take on the risks inherent in a contract should be employed. Insolvency usually terminates a company's insurance cover, so any liabilities discovered will not be claimable and may return to your organisation.
- Very often, your organisation will amend or add to the standard forms of contract. They may even have bespoke contracts. Make sure that you understand the purpose behind such amendments and how they are intended to be applied.
- Your company may be a party to a contract which is not one of those promoted by the ICE. It seems reasonable for a candidate for ICE membership to know why another form of contract is preferred.
- Understand fully the way in which the risks in a contract have been apportioned. You may, for example, believe that the risk of adverse weather is solely the responsibility of the contractor, but it may be that, if the weather is particularly unusual or severe, the risk is the client's.
- Be aware of what remedial actions are available in the event of non-performance by a contractor. What rules are there for agreeing that work needs to be redone or modified? When and how can work be accepted which is to a lower standard than originally agreed?
- Damages are awarded as compensation for loss, not as punishment; the law states 'such damages as may fairly and reasonably be considered as flowing naturally from the breach'. Liquidated damages are applied where the damages for a particular breach of contract have been agreed by the parties in advance, typically for late completion (for example, the value of the trade lost by late opening of a supermarket). Penalty clauses are unenforceable in English law, but are still occasionally referred to by the ill-informed.
- There is duty on a party to minimise loss resulting from a breach by the other party. Where an event arises which gives a party the right to a claim or compensation event, they have a duty to keep the costs to the employer to a minimum. Any damages awarded will be reduced by any amount which is needlessly spent.

As a professional civil engineer, it is vital that you read, discuss and understand every contract of which you are part. Never wait until something goes wrong before finding out what rights and responsibilities you have, and what you and your employer can properly expect of the other party.

Initial Professional Development for Civil Engineers
ISBN 978-0-7277-4147-9

doi: 10.1680/ipdce.41479.077

Chapter 11
Health, safety and welfare

Institution framework

The ICE offers the following descriptors in Appendix A of ICE3001 for this attribute.

> **6. Health, Safety and Welfare**
> A **sound knowledge** of legislation, hazards and safe systems of work
> Ability to manage risks
> Ability to manage health, safety and welfare within own area of responsibility

In addition, for Chartered status:

> Commitment to **leading** continuous improvement in health, safety and welfare

Superficially, these few words do not appear to cover very much. But the implications are onerous, as can be seen when the Development Objectives are scrutinised:

> **E2 Manage and apply safe systems of work**
> - Identify and take responsibility for own obligations for health, safety and welfare (HS&W)
> - Implement HS&W systems
> - Current HS&W codes and legislation
> - CDM Regulations
> - Risk assessments
> - Method statements
> - Recommend improvements
> - Recognise compliance costs
> - Detailed knowledge of hazards in own field of work
> - Appropriate risk management techniques
> - Safety briefings and inductions
> - Prepare and review risk assessments and method statements
> - Proactive approach to HS&W practice and management

With the addition, for Chartered status, of

> Contribute to the development and improvement of systems

Membership Guidance Note 20 explains what is required to satisfy the Review standard for health, safety and welfare, and gives some guidance on how it might be achieved and demonstrated.

MGNs 21 and 22 seek understanding of the following topic areas:

> Health and safety management
> Health, safety and welfare during the design construction, maintenance, operation and subsequent removal of the works
> Quality, health and safety and environmental management systems

Health, safety and welfare

Worldwide, social and professional attitudes to health and safety are reflective of differing cultures' particular attitudes towards the value of human life. It is important that anyone operating outside the UK but wishing to be a member of the ICE has an understanding of UK law and practice, in order to be clear about any differences between UK practice and their personal experience in different cultures.

In the UK, health and safety (welfare is a relatively new item although, tacitly, it should always have been there) is not just about such basic things as personal protective clothing, toe boards and completing standard forms for risk assessment. The superficial emphasis on such details, prompted by the need for written evidence of compliance to combat an increasingly litigious attitude in society, has inevitably drawn attention away from the fundamental concepts to the details of following the rules.

Recent publications from the ICE and from the Health and Safety Executive (H&SE) are reinforcing a 'back to basics' approach. The H&SE website (which is a mine of clear, useful information, much of which can be downloaded free of charge) states:

> If you believe some stories you hear, health and safety is all about stopping any activity that might possibly lead to harm.... This is not our vision of sensible health and safety. Our approach is to seek a balance between the unachievable aim of absolute safety and the kind of poor management of risks that damages lives and the economy.... In a nutshell – risk management – not risk elimination. (www.hse.gov.uk)

The health, safety and welfare of everyone, not just constructors, must be the bedrock upon which all decisions are made. Together, they constitute an attitude of mind, well beyond the humdrum details. That attitude requires

- a responsible and relevant belief in the sanctity of human life
- the vision to anticipate what could happen at every stage of the procurement, use, maintenance, modification and demolition of infrastructure

- the ability to make sensible judgements on what is likely to be acceptable to that particular society at that particular stage in its development.

These onerous and ever-changing requirements require many more skills and much more understanding than just technical judgement or compliance with an established system of risk management.

Increasingly, knee-jerk reactions to accidents in the UK have led to a reactive waste of resources. For example, was the nationwide increase in the length of motorway safety barriers in the UK justified because one driver fell asleep, drove across a field and landed on a railway line, causing a horrendous rail crash? Was that the best risk mitigation or avoidance? Perhaps politically it was, since the government had to 'be seen to be doing something'. But could there perhaps be more effective alternatives? Would tachometers have been a better mitigation measure? Candidates for Review will be expected to be thinking about safety problems in this broad manner even though, at this stage in their careers, they may not be in a position to influence the outcomes.

In March 2005, there was an industrial accident at the BP Texas City refinery, in which 15 persons were killed and some 170 seriously injured. The independent Baker inquiry into the catastrophe stated something of which we should all continually remind ourselves:

> Preventing... accidents requires vigilance. The passing of time without [an accident] is not necessarily an indication that all is well and may contribute to a growing and dangerous sense of complacency. When people lose an appreciation of how their safety systems were intended to work, safety systems and controls can deteriorate, lessons can be forgotten, and hazards and deviations from safe operating procedures can be accepted. Workers and supervisors can increasingly rely on how things were done before, rather than rely on sound engineering principles and other controls. People can forget to be afraid. (Panel statement, *The Report of the BP US Refineries Independent Safety Review Panel*, 2007)

The provision of protective equipment (such as high visibility clothing, protective footwear or safety harnesses) can foster a sense of complacency, where operatives no longer feel unsafe and so take unnecessary risks. As an example, I have moved workers who were lunching with their backs against the concrete safety barriers on a motorway. Such equipment is not intended as a substitute for continuing vigilance, but as a back-up should that vigilance prove insufficient. Engineers must continually remind everyone of their responsibility to protect the 'safety of themselves and those around them' (Health and Safety at Work etc. Act 1974), to the extent that safety equipment should never be needed.

Engineers must assess intelligent, thoughtful and holistic protection and judge whether it is appropriate for the circumstances. It is a matter of constantly assessing ever-changing criteria in situations where there can never be universally 'correct' answers. But, as engineers, we must always be able to prove that the measures we have in place are sensible, appropriate, practical and economic.

Forms and procedures record and communicate this fundamental attitude to life and limb, but must never be seen as an end in themselves, rather as a means by which the responsibilities are discharged. Technicians and operatives must know and comply with these processes, but need continual reminders not to 'forget to be afraid'.

Professional engineers, particularly those aspiring to Chartered status, must never see compliance as sufficient in itself, but simply as a means of recording how their responsibilities have been properly discharged. Their role is to have the vision and foresight to anticipate what could happen under all kinds of conditions, and then to determine how any hazards could be avoided or, at the very least, how the risks can be minimised. This is an ongoing responsibility, which cannot be fully discharged at the start of any project. As the project moves forward, the risks change and need to be reassessed, sometimes on a daily, even hourly, basis.

So, to start right from the basics, just what are health, safety and welfare?

Health

The World Health Organisation (1993) tells us that:

> health comprises those aspects of human health, including quality of life, that are determined by physical, chemical, biological, social and psychosocial factors in the environment.

This is a much wider description than most seem to realise. It is a bit wordy, but does go on to spell out our overall responsibilities, and the need for judgements, very clearly:

> [It is] the theory and practice of assessing, correcting, controlling and preventing those factors in the environment that can potentially affect adversely the health of present and future generations.

Not a bad description of the role of the professional civil engineer.

Safety

This is rather more difficult to define; there is no 'standard'. Every society, every culture, indeed every individual within those groups, will have a different perception of what is 'safe'. Some time ago, I would have been the first to slide on the ice or get on a fairground ride. Now I fear that I might fall awkwardly or feel sick. My attitude towards safety (my 'risk management') is changing with age. Young children have a propensity to engage in hazardous activities, because they have not yet realised, or been told, the risks. In the UK, some parents and guardians have reacted to this by attempting to avoid all risks to their children, who are said to have become unadventurous as a consequence.

There is an expectation, apparent in large sections of the UK public, that somehow all risk can be avoided and, if it is not, then someone else must be to blame – the 'compensation culture'. There are people who truly believe that the railways should never have a derailment. People are concerned about the risks of travelling by air or rail, but think nothing of driving on UK motorways which have a far greater accident risk, with 3000 people killed annually. Public perception is fickle, and engineers must avoid being sucked into irrational reactions.

Safety is about *managing* risks, not trying to eliminate them. The HSE states that:

> the Health and Safety at Work Act does not require firms to obey inflexible, hard and fast rules, rather they need to assess the risks that result from the work and identify sensible control measures that are proportionate to the risks. (Walker, 2005)

So, I believe that all candidates for membership must demonstrate that they have the necessary vision to anticipate hazards and the capability to mitigate risks in investigation, design, construction, use, maintenance and demolition, within criteria defined in law or established as best practice.

Obviously, the bias will be towards those parts of civil engineering in which you have direct involvement. While, for a potential Technician Member, it may well be mainly about compliance with procedures and systems, potential Members are required to demonstrate that they are capable of using judgement in achieving acceptable solutions and are prepared to defend their decisions, not merely comply with rules in an attempt to avoid blame and possible prosecution.

Safety culture

Most civil engineering organisations have a commitment to a safety culture throughout their operations. The intent is to develop, foster and maintain an attitude, both collectively and individually, which encourages safe working practices. Such a policy has to be actively and continuously pursued through what can be summarised as the 'four Cs'

- control – monitoring, auditing, supervision
- cooperation – ensuring participation and commitment
- communication – signal commitment, meetings, 'toolbox talks', posters, bulletins
- competence – training, certification.

You must know what your employer is doing to promote the safety of its workforce and all the people affected by its work, and have views on the effectiveness of those procedures.

Welfare

This is an even more nebulous concept, the 'faring well' of everyone involved. The benefits cannot be so easily demonstrated. Safety (or the lack of it) can perhaps be 'measured' by listing accidents and health by changes in the number of days off work, but the benefits of welfare are not so clear. Yet all three are interrelated, particularly when there is a serious attempt to develop a 'safety culture'. The provision of adequate and decent washrooms, toilets and catering facilities, well-decorated and spacious offices or comfortable safety and weatherproof wear which allows freedom of movement do not, of themselves, improve health and safety, but they do help to create an overall attitude. Buildings with few external windows and inadequate lighting and ventilation do tend to produce low morale, higher staff turnover and absenteeism, contributing to the term 'sick building syndrome'. Those who have experienced the noise of working next to highway contra-flows will understand the sanctuary of soundproofed accommodation, not itself a safety measure, but surely contributing something to a safer working environment by reducing stress?

Welfare is not restricted to site work but affects all aspects of infrastructure development, use, maintenance and demolition. Take note of what is being done around you, what you could do or suggest to provide a better work environment, and demonstrate your concern to your Reviewers.

The legislation

The base legislation is the Health and Safety at Work etc. Act, enacted in 1974. From this has come a host of regulations, which apply in the majority of work situations. You must find out which affect your circumstances and make notes of how they were applied. Add the dates they first came into force and any subsequent amendments. Many organisations have a technical library of such documents, rigorously kept up to date, so use it.

Particularly noteworthy are the Construction (Design and Management) Regulations, covering safe systems of work on construction sites and, subsequently, the Management of Health and Safety at Work Regulations, which cover risk assessments (see below). Other regulations cover such areas as ventilation, heating, lighting, workstations, VDUs, seating and welfare facilities.

Appropriate protective clothing and equipment for employees (personal protective equipment – PPE) is covered by the Personal Protective Equipment at Work Regulations, while restrictions on moving heavy objects by hand are contained in the Manual

Handling Operations Regulations. Control of Substances Hazardous to Health Regulations (COSHH) require employers to assess the risks from hazardous substances and take appropriate precautions. There are specific regulations covering particular hazards (e.g. asbestos, lead, explosives, gases and chemicals).

The list above is by no means exhaustive. Only you, in your particular circumstances, can compile a complete list of all those pieces of legislation which affect you. List them, know how your employer complies with them and make notes.

Risk analysis and risk management

Risk analysis is a process of carefully examining what could cause harm to people. Not all of these hazards can be eliminated. For those that cannot, then risk assessment involves deciding what precautions should be taken to prevent or minimise harm. The two, together, form risk management.

A 'hazard' is anything with the potential to cause harm (e.g. chemicals, machinery, working at height or alongside a busy highway – even the weather).

A 'risk' is the chance that someone will be harmed by the hazard. The risk is a function of two things

- the possible severity of the harm
- the number of people who may be harmed.

For example, the risk in a busy urban area of children being killed by traffic is quite high. Limiting the traffic speed (to 30 mph) has been shown to save lives and is widely adopted. Lower limits (20 mph) can reduce the risk of serious injury. Similarly, traffic speeds are reduced past roadworks to limit the risk of injury. But, in addition, substantial protective barriers between vehicles and the workforce are required in the UK.

Accidents can ruin the lives of entire families. Those in positions of authority, such as professional engineers, may not have been proved negligent or foolhardy but have to live with the moral dilemma of wondering whether, after an accident, there was not *something* they could have done to further reduce or even avoid the accident.

Every accident affects the business directly – through delays, lost output and damaged equipment – while serious incidents may involve fines and legal fees and will increase insurance costs. Court cases consume inordinate amounts of staff time. Adverse publicity can affect the company's reputation and its ability to win work.

The effort and detail that is required to prepare a risk assessment depend on

- the level of risk
- the nature of the work
- the complexity of the workplace
- the type of people involved.

When first introduced in the UK, risk assessments tried to identify every possible hazard and assess every possible risk. The inclusion of relatively trivial risks, of which properly recruited and trained workers involved were well aware, could and did obscure less obvious risks which, not being drawn specifically to anyone's attention, were therefore more likely to happen.

Risk assessments must be relevant to the task, area, work and persons that may be present, identifying only unusual risks beyond those which are expected by an adequately trained workforce. Since they cannot be briefed or trained, risks to the general public must be avoided.

Compiling risk assessments

Risk assessment is a serious responsibility, routinely undertaken by civil engineers. Its success is dependent on several personal qualities, including

- the vision to anticipate what might happen
- experience and knowledge of the kind of location, tasks and hazards involved
- the ability to make rational and realistic judgements on the degree of risk
- a knowledge of available safety equipment and methods which may assist in mitigating any residual risks.

Any assessment should involve all or some of the persons likely to be exposed to the risks. They will have their own specific experiences which may usefully add to the assessment, but mainly their inclusion is to ensure that staff and workforce accept ('take ownership of') any control measures and implement the correct procedures without excessive supervision. However, the responsibility for correct implementation remains with the risk assessor.

Completing risk assessments

The key piece of UK legislation is the Management of Health and Safety at Work Regulations, introduced at the beginning of 1993 to comply with a European Directive. The Regulations require that:

> Every employer shall make a suitable and sufficient assessment of
>
> - the risks to the health and safety of his employees to which they are exposed whilst they are at work; and

- the risks to the health and safety of persons not in his employment [including the public, subcontractors and visitors] arising out of or in connection with [the employer's work].

The Regulations go on to state that:

Every self-employed person shall make a suitable and sufficient assessment of

- the risks to his own health and safety to which he is exposed whilst at work, and
- the risks to the health and safety of other persons not in his employment arising out of or in connection with [his work].

There is sometimes a mistaken view that the employers alone are responsible for the health and safety of their employees. Each employee has responsibilities similar to those of self-employed persons and must not rely solely on their employer for protection or, worse, do anything foolhardy or irresponsible. However, personal assessments are rarely written down or recorded. This is where a written record of adequate workforce training is advisable.

Most organisations have standard forms for recording risk assessments. You must understand your procedures, remembering that forms are not substitutes either for proper identification of all significant hazards or for the application of sound judgement in mitigation of the inherent risks.

Risk assessments must be reviewed routinely whenever working practices, the people employed or the working environment change significantly, or following an accident or 'near miss'. This requires constant vigilance and awareness on the part of civil engineers and others to monitor what is happening and realise the possible changes in hazards and/or risks and act accordingly.

There are five stages in assessing risks, which are outlined below.

Stage 1 – Hazard identification

The hazards should be identified by

- walking around the workplace and determining what could reasonably be expected to cause harm (concentrating on the significant hazards that could result in harm to one or more people) in both the short and long term
- asking the workforce for their views (they may have noticed things which are not immediately obvious)
- using equipment suppliers' instructions, material safety data sheets, etc.
- using industry guidance and regulatory information
- referring to information from clients or third parties when working on other sites.

Stage 2 – Decide who might be harmed and how

This includes all employees but, particularly, vulnerable staff such as young or new workers. In offices and public places, it includes the effects on visitors, external contractors, maintenance workers and members of the public. Never forget persons with disabilities, non-native speakers or the deaf, expectant or new mothers and children. Judgements have to be made on how and when people might be exposed to the hazards and potentially harmed.

Stage 3 – Risk evaluation

Evaluate the risks and decide whether existing precautions are adequate or whether more should be done. At this stage, the risk of each hazard causing harm to someone is put into some sort of hierarchy, focusing on eliminating significant risks or reducing the level of risk to a minimum but discarding trivial items.

Risk reduction can be an expensive process, so the law allows a 'reasonably practicable' approach. This means that costs, time and effort can reasonably be taken into account to decide what control measures to implement; for example, it could be 'reasonably practicable' to send one member of staff on an external training course, who then passes on their knowledge to colleagues in the form of team meetings or toolbox talks, as long as there are written records. You should know how this 'reasonably practicable' approach is interpreted in your organisation.

The decision of whether risks are high, medium or low is very subjective. Everyone has a different perception depending on their experience, knowledge, training and, indeed, age. Your organisation will have defined a consistent approach to calculating levels of risk and you should know what this is and how to apply it.

To control risks, the following principles are applied

- try to avoid the hazard all together
- tackle the risk at source
- if the hazard cannot be eliminated, reduce the risks
- adapt the work to the individual rather than expecting people to fit the job (allow staff to work within their capability)
- provide appropriate instruction and training and document this on training records.

At the end of this process there will inevitably be some risks remaining. Know how an action plan is compiled, prioritised on the criteria outlined in the stages above, with a realistic timetable for implementation.

Other control measures may include health surveillance for specific hazards (e.g. working with hazardous chemicals, noise, vibration or asbestos). Take note of those you come across.

Stage 4 – Record the findings

It is recommended that *all* organisations document their assessments, irrespective of the number of employees, even though five is the official minimum. Employees must be given time to read and understand the findings of risk assessments and have the opportunity to comment. Once understood, the assessment should be signed. There will be a system for bringing the attention of any visitors or subcontractors to relevant assessments, so find out how it works.

Written records must be available during routine inspections or accident investigation to show that hazards and associated risks have been adequately assessed. In the event of actions for civil liability, good records can demonstrate that reasonable steps had been taken to prevent harm.

Stage 5 – Review the assessments and revise if necessary

Assessments should be reviewed regularly, at least annually but more frequently for high risks, to ensure that the precautions are still working effectively and that familiarity is not breeding complacency. New equipment or machines, substances, procedures or people may lead to new hazards which should be assessed before their introduction.

Risk assessment format

There is no legally required format for undertaking risk assessments; they should be in a simple format which is readily understood by employees. For example, Citation plc (a health and safety consultant) has a web-based general risk assessment system called Citassess, based on using simple picture icons to demonstrate hazards and controls associated with a particular risk. Others use a simple paper-based 'matrix style' assessment. Find out why your organisation uses its chosen system and discuss how effective it is. Think about possible improvements in format, style and delivery.

Summary

In a nutshell, do you fully understand all the possible consequences of your work?

Those applying for Chartered Membership must also demonstrate that they are using their vision and initiative to strive for continuous improvements in HS&W. This does not necessarily require you to be in a position of authority; we can all influence attitudes and behaviour by example, or by appropriate discussion and criticism. Demonstrate that you can!

Initial Professional Development for Civil Engineers
ISBN 978-0-7277-4147-9

doi: 10.1680/ipdce.41479.089

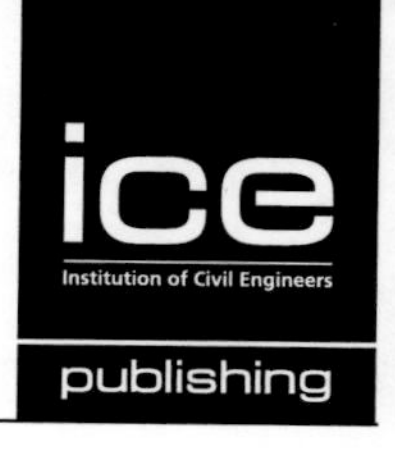

Chapter 12
Sustainable development

Professional civil engineers are increasingly being required to play a leadership role in sustainable development, overcoming global challenges caused by such things as depletion of resources, environmental pollution, climate change, rapid population growth and damage to ecosystems. Such major and fundamental considerations can seemingly be beyond the spectrum for recently graduated engineers, but are of inestimable importance to our work and its influence on the world. Sustainability must not, and cannot, therefore be overlooked by any professional civil engineer. It will inevitably become an intrinsic part of all our work.

Institution framework

The Institution offers the following descriptors for this attribute in Appendix A of *ICE3001 Routes to Membership*:

> Sound knowledge of sustainable development best practice
> Manage engineering activities that contribute to sustainable development

In addition, for Chartered Membership:

> Committed to **leading** continuous improvement

The Development Objectives offer:

> **E3 contribute to sustainable development through engineering activities**
> - Comply with legislation and codes
> - Environmental impact
> - Environmental management
>   - Interaction of design, construction, operation, decommissioning, demolition and disposal
> - Social and economic issues in implementing solutions
> - Project whole life cycles
> - Knowledge of sustainable development tools

The topic areas listed in the MGNs 21 and 22 include:

Environmental issues in civil engineering Environmental management systems National and international regulations on the control of pollution

MGN 47 explains how candidates for Professional Review might demonstrate their commitment to sustainability.

Professional framework

The ICE Royal Charter (1975) requires 'knowledge and judgement in the use of scarce resources, care for the environment and in the interests of public health and safety'. The Code of Professional Conduct requires us all (Rule 4) to 'show due regard for the environment and for the sustainable management of natural resources'.

In the UK, the Engineering Council produced a booklet in 1993 *Engineers and the Environment*. This was updated in 2009, when six principles were stated, not only to guide and motivate members to achieve sustainable development in their work, but also to meet their professional obligation to seek sustainability. Fully compatible with UK-SPEC and the UK government's Sustainable Development Strategy *Securing the Future*, the six principles are:

(*a*) Contribute to building a sustainable society, present and future
(*b*) Apply professional and responsible judgement and take a leadership role
(*c*) Do more than just comply with legislation and codes
(*d*) Use resources efficiently and effectively
(*e*) Seek multiple views to solve sustainability challenges
(*f*) Manage risk to minimise adverse impact to people or the environment.

The full document can be downloaded from the Engineering Council's website and should be read alongside sustainability related information produced by our Institution, such as the SAID Report – *Sustainability and Acceptability in Infrastructure Development*, originally written in response to a challenge from the then Environment Minister, John Prescott.

Historical background

This book cannot cover every aspect of the growth of the desire for a more sustainable future. What follows covers some of the important world events leading up to the present position. Many, if not all, civil engineers will be influenced and guided by regulations coming from world conferences and particularly the EU, which have resulted in the development of company systems with which you will be expected to comply. To gain a proper understanding of the purpose of the rules, it is important that you are able to relate them to their origins. This chapter should help you in that endeavour.

As the world population continues to grow, the demand for energy and resources increases. Developing nations aspire to reach or exceed the material prosperity enjoyed by the developed world. Natural resources are being consumed at a rate which many predict will impoverish generations to come and result in serious climate changes. The world is beginning to realise that these trends cannot continue unchecked.

Sustainability is not a new concept. As long ago as 1789, Thomas Jefferson, the third American president, is recorded as saying, 'No generation can contract debts greater than may be paid during the course of its own existence.' But very few people heeded his prophetic words. It is really only since the 1960s that politicians have been made aware, by increasing activist pressure, of the consequences both of continuing to plunder non-renewable resources and of pollution. By 1972, this increasing awareness caused the United Nations (UN) to hold a major Conference in Stockholm on the Human Environment. As was to be expected, the developing countries claimed that the environment is being despoiled mainly by the advanced industrial nations, whose belated environmental concerns could hamper their own development. In contrast, the developed nations were concerned that the imposition of enhanced environmental standards will restrict their economic growth and erode their standards of living.

The World Commission on Environment and Development (generally known by the name of its Chairman Harlem Brundtland) was convened by the UN in 1983 to address growing concern 'about the accelerating deterioration of the human environment and natural resources and the consequences of that deterioration for economic and social development'. The report by the Brundtland Commission, *Our Common Future*, gave a definition of sustainable development which has become familiar:

> Sustainable development meets the needs of the present without compromising the ability of future generations to meet their own needs.

This definition contains two key concepts

- the concept of 'needs', in particular the essential needs of the world's poor, to which overriding priority should be given in preference to the 'wants' and 'selfish desires' of the already wealthy and
- the idea of limitations imposed by the state of technology and social organisation on the environment's ability to meet present and future needs.

Ongoing efforts by the UN to find solutions to perceived problems, including pollution of the Mediterranean Sea, the threat to aquatic resources, deforestation, desertification, depletion of the ozone layer and global warming, led to the realisation that reconciling the goals of environmental protection and continuing development is far from easy and any agreements would take a long time to broker. This impasse eventually led to the UN Conference on Environment and Development in 1992 in Rio de Janeiro, by far the largest intergovernmental conference until then. Known as the 'Earth Summit', it produced a declaration of principles (the Declaration on Environment and Development), a plan for sustainable development of the Earth's resources (Agenda 21) and guidelines for the sustainable development of forests, generating a surge of commitment which has continued to influence subsequent policies.

In 1997, after much debate, a unique international initiative recognising the particular severity of the problems caused by global warming was a protocol agreed in Kyoto, Japan. Its intention is to limit the emissions of the four most damaging greenhouse gases (carbon dioxide, methane, hydro fluorocarbons and sulphur hexafluoride). The Kyoto Protocol came into force in February 2005, and has so far been signed by about two-thirds (187 in 2009) of the world's nations. The notable exception is the United States, responsible for about 25% of global emissions but, at the time of writing, the current President, Barack Obama, has suggested that this may change.

Another attempt by the UN to reinvigorate debate on sustainable development was made in September 2002 in Johannesburg. This conference recognised that, in the ten years since the Earth Summit, world poverty had deepened and the environment had degraded further. There was a general commitment to new targets, with timetables, but with the pragmatic message that there could be 'no magic and no miracles – only the realisation that practical and sustained steps were needed to tackle many of the world's most pressing problems'. Possibly the most far-reaching outcome was the realisation that poorer countries were going to need resources from the wealthy nations to support any sustainable development (i.e. there would have to be a significant redistribution of wealth).

The EU has been an enthusiastic promoter of sustainability since the early 1970s and significant resources have been devoted to conferences and committees drawing up increasingly tight controls on such things as biodiversity and greenhouse gases (notably carbon dioxide CO_2) emissions. This overview cannot possibly list all the almost yearly environmental reviews since the EU ratified the Kyoto Treaty in 2002, or the policy statements which have resulted. It is highly likely that some of these conferences and commissions spawned the rules to which you are required to adhere in your daily work; you should do your own research to link each of them to their source.

Perhaps the most noticeable and far-reaching effect of EU legislation in the UK so far came in 1992, when the European Court of Justice ruled that the UK government was in breach of a 1976 directive on water quality, notably regarding the bathing beaches around Blackpool. Drinking water also had levels of nitrates in contravention of a 1985 Directive. The Mersey earned the title of the dirtiest river in Europe. These failures led to a major investment in waste water collection and more sophisticated treatment, such that the UK now has some of the least polluted coastal waters in Europe and fish and other aquatic life are now abundant in the Mersey and other rivers.

Practical sustainability

Sustainability is an ideal. Satisfying the needs of today without compromising the ability of future generations to meet their needs is not possible all the time we continue to use non-renewable fossil fuels and dig ores and aggregates out of the Earth. The Industrial Revolution and its aftermath have put paid to any idea that those of us in the developed world can quickly or easily return to living with, rather than on, the Earth. Only some ancient civilisations, which have not joined the industrial surge, could possibly hope to do that. Having spent a few days with the Native Americans in the desert, I came away with a profound sense that we had much to learn from them, but such major changes as are necessary will be strongly resisted as retrograde by the majority of the developed world.

Sustainable development is the pragmatic response to the ideal concept of sustainability. Sustainable development is the management of resources in a project to maximise the benefit while minimising the disbenefits and damage to the environment. For this purpose, the best definition of the environment is, perhaps, the aggregate of all the external conditions and influences affecting all forms of life on this planet, both now and in the future.

The idealistic concept does seem to suggest that sustainability only applies to the human race ('future generations') but I feel sure this is not what was intended by the World Commission. Hence, the broad description of the environment above, covering many aspects beyond the human race.

We can, and must, work towards the ideal, and everyone approaching the Reviews must demonstrate their awareness of current *best practice* and how it is implemented in their work. It is unlikely to be enough to show that you avoided landfill tax by re-using materials within the confines of the site, or crushed some redundant concrete to form fill (at what energy cost?). Sustainable development is far more fundamental than that, requiring a thorough understanding of the full complexities of infrastructure development.

In the past, infrastructure development was a progression through a series of steps: market analysis, land acquisition, engineering and financial constraints, conceptual planning, engineering design, construction, modification. In that progression, requirements such as energy, transport, water supply and wastes removal, tended to each be considered individually, while social and cultural elements were often overlooked or perceived as obstructions.

Now the design process starts with gaining an understanding of the natural ecology of a site: topography, geology, hydrogeology, soil types, climate, habitat, flora and fauna. By building a comprehensive inventory of the existing environment, patterns emerge, suggesting that some areas are more suitable for the proposed development than others, and also requiring the development to be much more sympathetic to what is already there.

Much of this newer thinking is now embedded in most organisations' systems of work. For many organisations, the 'bottom line' now has three components – profit, people and planet, or the three pillars of sustainability. There is therefore a chance that you may take sustainability for granted, not realise the full significance of what is being done and subsequently fail to demonstrate your understanding at Review. So do consider what is behind the rules and systems that you use.

The Institution too, has produced guidance; have you read it and sought to put it into practice? If so, tell the Reviewers about it. Candidates for the Chartered Review must additionally demonstrate that they are continually seeking every opportunity to make inroads into further reducing our use of non-renewable resources. In other words, that they are complying with the Engineering Council's principle to 'do more than just comply with legislation and codes'.

One of the key aspects of sustainable development is the realisation that sustainability requires that 'needs' are met, not, as is so often the case in developed societies, selfish 'wants' or 'desires'. So you must develop an awareness of any efforts made, of which you may perhaps be an unwitting part, to persuade the public to reduce their desires, which are heavily influenced by massive retail advertising in our developed society. For many years, engineers in the UK and elsewhere anticipated demand and attempted to provide for it ('predict and provide'). Now we are beginning to restrict such demands and, inevitably, facing a backlash of complaints from a public used to getting what it believes it wants.

At Review, a sound knowledge and commitment to sustainable development can be demonstrated by showing that you and your employer exploit any chances to reduce environmental damage. This could include such everyday things as

- fitness for purpose – neither grandiose nor over-demanding over the life cycle
- design and construction which at least protects the existing ecology and possibly improves it
- reclamation, recycling and re-use
- reducing specifications to allow secondary materials to be used where safety and fitness for purpose are not compromised
- refit rather than rebuild
- energy conservation in the entire life cycle of a project.

Environmentally acceptable

As public perception changes so, too, do the demands made on civil engineers. We are expected to anticipate what the public's attitudes are going to be at the time any project is implemented, which may be years away. It is no longer possible to impose solutions without adequate explanation and the development of a consensual public view. But, at the same time, we must balance 'needs' against 'wants' by convincing a sceptical public that our engineering judgement is of the highest order – and correct, socially, environmentally and politically, at the time that decisions are made.

Environmental Impact Assessment (EIA)

The Department of the Environment defined EIA as a

> systematic technique for drawing together expert quantitative analysis and qualitative assessment of a project's environmental effects, presenting the findings in a way which enables the importance of the predicted effects, and the scope for mitigating or modifying them, to be properly evaluated. (Circular 15/88)

EIA is a complex process, logically going through all aspects of both the existing and potential situations. Several best practice methodologies have been developed, all incorporating similar basic steps.

(*a*) Identify the problem – this will encompass many diverse aspects and their interrelationships, such as landscape and land use, including visual impact, geology and topography, hydrology and groundwater, air quality, ecology of flora and fauna, noise, traffic and vibration, antiquities and archaeology, socio-economics, cultural heritage, communities, existing infrastructure, amenities, travel patterns, travel modes and journey times.
(*b*) Predict the impact of the proposal on each aspect in the above list and anything else.
(*c*) Evaluate the significance, positive and negative, of that impact.
(*d*) Devise strategies to mitigate, wherever possible, the negative effects.
(*e*) Formally present the findings, usually in the form of a Report.
(*f*) Review public and official political reaction.
(*g*) Decide and deliver.

At appropriate stages in the process, it is necessary to gain the knowledge and understanding of all those affected, through consultation and participatory meetings. This requires careful decisions on who should be consulted, when and how. I have personal experience of great success, if this consultation is done early, comprehensively and openly, such that when originally contentious proposals were formally put to the planning authorities, no formal objections were lodged. But it can take years and considerable (and costly) effort.

Everything civil engineers do impacts on the environment, both in the vicinity of the proposal and in sourcing and moving the materials needed. So every civil engineer must be aware of the possible ramifications of their proposals. Even such relatively simple issues as the choice between concrete, asphalt and porous road-wearing courses are not dictated purely by economics, but embrace such things as sourcing materials, vehicle noise, spray, fuel consumption, maintenance and useful life. The EU imposed an end to the dumping of sewage sludge at sea by the end of 1988 (EU Urban Wastewater Directive 91/27/1/EC): but is incinerating sewage sludge, with its energy use and possible air pollution, better than potential ground or seawater pollution? Could conditions be realistically achieved where the sludge naturally biodegrades, creating gases as possible pollutants? These are difficult judgements, but civil engineers must make them and 'sell' them to a sceptical public.

Part of the drive for more environmentally friendly infrastructure is the consideration of whole life costs and value engineering: topics covered in greater detail in Chapter 8.

Initial Professional Development for Civil Engineers
ISBN 978-0-7277-4147-9

doi: 10.1680/ipdce.41479.097

Chapter 13
Interpersonal skills

It is ironic that, at the tender age of 13 or 14, many of us in the UK made educational decisions to 'go on the science side' in preference to the humanities and English, and then found that these discards were vital if we wanted to become competent engineers. This book is not the place for detailed discussion of the use of English, grammar and syntax or the style of reports; there are plenty of specialist publications on all these aspects. Here we are discussing the basic principles of communication.

Institution framework

The Institution offers the following descriptors for this attribute:

Communicate well with others at all levels
Discuss ideas and plans competently and with confidence
Personal and social skills

and additionally for Chartered status

Communicate new concepts and ideas to **technical and non-technical colleagues**

Elsewhere, in the Development Objectives, the Institution mentions:

D1 Communicate with others

- Reports, letters, drawings, etc.
- Presentations
- Exchange of information
- Advice to technical and non-technical colleagues
- Contribute to meetings
- Consider the views of others
- Consultation
- Present your case and defend it
- Conduct and lead discussions

D2 Personal and social skills
- Awareness of needs and concerns of others
- Develop good working relationships to achieve collective goals
- Equal opportunities and diversity
- Set an example for others to follow
- Identify and agree collective goals

and in the MGNs 21 and 22 for the Written Test or Assignment:

Knowledge transfer
Effective delegation
The professional development of civil engineers
Training and development of staff

This is a lengthy list, which does emphasise the huge importance of communication in civil engineering.

Principles of communication

Communication is about the transmission of accurate and concise information. Information is not just facts, but covers things such as support, encouragement, discipline, attitudes, policy, goals, vision, ambitions, even dreams. We explain and complain (communicate) to achieve a purpose.

For most professional communication, the tone of transmission should be calm, factual and straightforward. It should also be respectful and courteous, which requires the sender to know and understand the social and cultural background of the receiver. There are many dangers here, such as

- underestimating the knowledge of the recipient, which is demeaning
- overestimating their knowledge, which just baffles them
- inadvertently offending established customs or strongly held beliefs
- imposing an overwhelmingly detailed explanation of all the factors, which merely antagonises them.

It is the consequence of the communication which is important, not the communication itself. The ease and availability of systems for communication is a threat to this fundamental concept. In former times, letters were the only long-distance means of communication. They had to be laboriously written and copied by hand and took days or even weeks to deliver. Great consideration had to be put into what and how much to say and how to say it concisely, to achieve the intended outcome. Much of today's communication is verbose, trivial, repetitive and banal, often seeking only to boost the image of the sender (as when telling someone off as a means of venting your own frustrations and anger) or to supposedly protect their reputation ('But I did tell you in an e-mail!' – which was a bland catch-all sent at the press of a button to anybody and everybody who might be remotely affected).

All of us are bombarded by ceaseless cascades of information, much of it irrelevant to us. As a result, it is all too easy to miss the one piece of information which is important. We, the receivers, are being forced to filter this plethora of information instead of, as should happen, the sender only sending that which is truly relevant *to us*.

Another major problem is the desire among certain people to actively distort and manage the message to achieve a clandestine purpose. They even employ specialists to do this, nicknamed 'spin doctors'. This technique, with its inevitable development of cynical mistrust, should have no place in civil engineering communication, where the reputation and status of our profession relies on provision of impartial, accurate advice based on sound understanding, without bias or prejudice.

Good communication is dependent on the effective relaying of just enough information to achieve the desired purpose. In too many communications, there is too much information. We have a tendency to include everything which we have thought about in compiling the message, most of which will be irrelevant to make the receiver act in the manner we desire. This seems particularly prevalent in report writing, with the result that few receivers (the clients) ever bother to read the whole thing. After all, the client has asked a question to which they do not know the answer; all they want is a substantiated answer which they believe. The detail should be recorded in the sender's Quality Assurance systems, for recall if the receiver requests further clarification. So do not include anything which is ancillary; if it does not help to achieve the purpose, leave it out. This is a fine balance – to achieve the trust that you do know what you are talking about, without information overload. It is the quality and content of the information which is absorbed by the receiver that determines the level of subsequent performance.

What is absorbed may not, indeed probably will not, be everything that is sent (tests reveal that a maximum of around 25% of a lecture is absorbed by the audience). So, it is important that the sender can monitor in some way how the information is being received. Obviously, face-to-face communication is in many circumstances the best way to do this but, even then, the sender must be acutely aware of body language and the things left unspoken by the recipient.

But this is not always possible, so frequently your assessment of the success of your message must be based on the outcome, the results achieved. This takes time and obviously precludes you intervening if the receiver goes off on the wrong tack. One good way to properly assess the effectiveness of your message is to ask the receiver, immediately after your delivery, to tell you what they intend to do as a result of the communication. This can be done both verbally and by writing and is a technique worthy of much greater use if misunderstandings are to be minimised.

Engineers gather, absorb, assess and process huge quantities of information and make complex judgements based upon it. 'At this time, with these resources, in these circumstances, for the foreseeable future, this is the best we can do' is a very strong engineering decision, but it is difficult to communicate because it is not a fact, but a judgement, based on many conflicting parameters. Engineers must issue the resulting conclusions in a manner which is understood by, and acceptable to, the recipients.

Our profession is generally not skilled at conveying our message, to other engineers, less to other professionals and even less to the public. There is a tendency to include all the base information from which we have drawn our conclusions, rather than just enough background for the recipient to be confident that we know what we are doing.

Our inability to communicate with the public effectively may be one of the reasons why our profession does not command the respect which we feel we deserve. Having thought the matter through, we tend to confront the public with the conclusion without adequately explaining the thought process behind it. Sometimes we present the solution before the public has even realised that there is a problem, let alone understood it or considered the options.

If the general public reacts negatively to a proposal, then it may well be our fault. The instigators of the project have not adequately explained why the proposal is the best solution to a given problem. Indeed, they may not even have properly explained the problem. It is not enough to arrive at the best solution and impose it. The solution must be accepted by the majority. Anyone who reasonably opposes the solution on the grounds of its adverse effects on them personally, must be offered adequate recompense. This could take the form of monetary compensation, relocation or accommodation works to significantly reduce those adverse effects. Inevitably, this will add to the project cost, but is a necessity if your solution is to succeed, and should be included in any project estimate. Any credibility which a pressure group has is the result of loopholes in our information supply (or the result of a mistaken judgement and underestimation of the resistance).

The process of communication involves a large number of decisions, many of which, in everyday conversation, are made subconsciously or without adequate thought. The flexibility and tolerance of normal personal relationships usually make adequate allowances for this.

But for communication, whether verbal or written, with strangers or organisations where personal relationships are not at that refined level, more thought is necessary. If the message is sent on behalf of the organisation, then it is the company's relationship with the receiving organisation which is important. Where the receiver is 'the public' then careful consideration is required, due to the diversity of individuals who will be affected by the message.

Medium

There are only three possible media

- speaking (e.g. face to face, by telephone, TV or video)
- writing (e.g. memorandum, letter, report, PR release, exhibition)
- visual (e.g. still or moving pictures, drawings, models, body language).

The many methods of transmission, such as post, fax, e-mail, video, Internet and satellite link, are only tools and cannot add to the absorption of the basic message. Indeed, all too often they detract from the message and the anticipated outcome is not achieved. In today's world of instant communication, it is all too easy to overwhelm the recipient with information. Do not include anything which is ancillary; if it does not help to achieve the purpose, leave it out.

Do not send unnecessary information to anyone from whom you do not need or expect a reaction. Copying several people into the same e-mail just so that they can see that you have done something is one typical example. Do they all need to know the same level of detail, or simply that you have done what was required?

Ease of use of a wide variety of communication systems is a constant danger to the quality of communication. Achieving a balance between all these variables is difficult and needs constant practice. Even among very experienced communicators, mistakes still happen. Arguably, more thought goes into the written word so, to discuss the basics, this medium is considered in detail. The same processes should be, but frequently are not, applied to speaking. Visual communication brings with it some additional problems, which are also discussed.

Written communication

You usually have no idea how or when a written communication is going to be received, and what emotions the recipient will be experiencing at that moment. So, face-to-face communication is undoubtedly the best medium, since you can watch the reaction and adjust your attitude and approach accordingly. Face-to-face, preferably in a closed, private environment, should always be used for bad news and disciplinaries. No one should hide behind the technology in an attempt to avoid the emotional turmoil of delivering bad news. Conversely, good news is often best delivered in public so that colleagues, friends or bystanders can enjoy the pleasure too.

Beware of being rushed into communication by the modern speed of communications – technology has vastly improved the latter, but the speed of thought of the human brain has not responded comparably. We have all, at some time, instantly regretted the too-rapid touch on the 'reply all' button! Indeed, some companies have altered their IT systems so that it is impossible to reply for 48 hours, in an attempt to force staff into thinking about the response – its content, timing, mode of delivery and, most importantly, the intended outcome.

Ideally, for *every* communication, whether verbal or in writing, the following questions need to be answered

- What is the purpose? What do we hope to achieve?
- What does the message need to be?
- Who is going to be the best sender?
- What is the best format?
- Which is the best medium for transmission?
- When should the message be delivered?
- Who is going to receive the message?
- What sort of language is appropriate?
- Might clarification be needed? Question and answer follow-up?
- What is the context in which the receiver will best take ownership, and so be able and willing to fulfil the purpose?

Dependent on the importance of the outcome, these questions all need a definitive answer *before* communication is attempted. Too often, too many of these points are overlooked, which is why so much communication is unsuccessful. Most, if not all, errors, mistakes and problems throughout procurement and management processes can be traced back to poor communication.

Quality assurance

Quality assurance (QA) is, in essence, a control system for communication, monitoring the process of relaying information. It provides an audit trail of what was done and said, by whom, when and why. QA rarely measures the effectiveness of what was absorbed by the receiver, but only records that the process of transmission was carried out to accepted standards. Few QA systems record whether the desired outcome was achieved or not (i.e. whether the communication was actually successful), unless it is something physical (such as a concrete pour or traffic lights), where the communication was a specification.

'Communicate well with others at all levels'

If you make a list of the people you, as an engineer, communicate with, even just on a regular basis, the one thing you will notice is the variety. One moment you may be talking to a client, the next to a workman digging a hole, to a piling foreman or the representative of a public authority, to another engineer or an aged householder. You may be required to write a letter to a disgruntled member of the public, an order for a subcontractor or a report for a government agency. Each of these has to be addressed differently, using different words, a different demeanour, different attitudes; each dependent on the knowledge and background of the recipient. The choices have to be made with care and consideration if you are to achieve a successful outcome. Your approach will significantly affect their response: done badly, your message will be rejected, whatever its merits; done well, your message will be received positively and, hopefully, acted upon.

The Transportation Security Administration, which looks after airport security in the USA, has enormous power over passengers, but takes interpersonal skills very seriously. To paraphrase the general definition of the attributes of a member of its personnel given on their website, they must consider and respond appropriately to the needs, feelings and capabilities of different people, in different situations; be tactful, compassionate and sensitive and treat others with respect.

Too many people are apparently interested only in projecting themselves, rather than feeling the audience's reaction; the transfer of information is in the wrong direction.

You must be able to give examples of the variety and, most importantly, the *effectiveness* of your communication skills to be successful at the Professional Reviews. You must be able to demonstrate, by examples, where you persuaded other persons that what you were proposing was reasonable and sensible, so that they cooperated fully in implementing the proposal.

This ability to persuade is particularly important for a potential Chartered Engineer, who must demonstrate that they can 'communicate new concepts and ideas to technical and non-technical colleagues' – that is, change the knowledge, understanding and attitudes of a whole range of varying recipients, by explanation and persuasion. You must demonstrate that you have such a capability.

Effective written or verbal communication is achieved by

- putting proper effort into research, delving into the background
- seeking help in defining the message (e.g. sourcing information and deciding how much of it is needed)
- ensuring the simplicity, accuracy and relevance of the information *to the recipient*
- appropriate timing – choosing the right moment
- correct environment, which will be conducive to reception
- best medium, something which is very dependent on the need to choose the best timing.

All these decisions take time and yet, all too often, we are in such a hurry to send the message that the thinking process, on how best to do it, is skimped.

Visual communication

With the huge strides in visual technology since the advent of computer graphics, there is now an amazing range of techniques available to display every project in all kinds of ways. Time and resources are the only limiting factors, just as for every project – 'at this time, with these resources, in these circumstances, for the foreseeable future', this is the best solution.

Yet, there is still a tendency to fall back on tried and tested techniques. For example, how many of the visitors at a public exhibition are able to imagine what a new road looks like from a helicopter, which is how the road was designed and drawn and, too often, how it is portrayed to the public? This visual communication is usual and convenient to the sender, but not to the recipient. So, straightaway, a barrier has been formed between the sender and the recipients. Only when the recipients can readily understand what is being presented will they tend to be more cooperative and not feel that they are 'being talked down to'. I have used watercolour paintings and photo-montages to successfully present proposals to the public. Nowadays similar pictures can readily be produced by computer.

Questions to encourage development of communication skills

Observe people around you whom you think communicate well and ask yourself the following questions:

- How do they achieve that consistently?
- Do they have a routine technique?
- Does their style vary from situation to situation?

What do people do which causes you to think that they are poor communicators?

List examples where poor communication has led to misunderstandings or mistakes and discover how they could have been avoided.

Where have you persuaded someone else to change their point of view, such as

- civil engineers
- other types of engineer
- specialists
- bureaucrats
- members of the public?

Consider times when other people have been very persuasive and consider how they achieved this.

Initial Professional Development for Civil Engineers
ISBN 978-0-7277-4147-9

doi: 10.1680/ipdce.41479.105

Chapter 14
Professional commitment

It could be said that this entire book is about the basic development of the attitudes, understanding and commitment of a professional civil engineer. So, this chapter is essentially a summary of the rest of the book in the context of our Code of Conduct.

Institution framework

The Institution offers the following descriptors for this attribute in ICE3001:

> Understanding and compliance with the ICE Code of Conduct
> Commitment to current and future CPD of self and others
> Support of ICE activities
> Personal commitment to professional standards, recognising obligations to society, the profession and the environment

The Development Objectives offer

> **E1 Comply with relevant code of conduct**
> - Purpose and history of ICE
> - Comply with ICE Code of Conduct
> - Ethical and professional behaviour
> - Current developments and issues affecting the construction industry
> - Promotion of the construction industry

> **E4 Manage your own CPD and assist others**
> - Review of your own development needs (DAP), record, evaluate outcomes (PDR)
> - Company appraisals

Everyone who applies to become a member of the Institution of Civil Engineers signs a declaration that they will, if elected, 'comply with relevant codes of conduct'. It makes sense therefore, for every civil engineer to know and understand the duties and responsibilities which that declaration imposes upon them – not merely technical competence, but legal, environmental, societal and financial as well.

ICE Code of Conduct

There are six rules in the ICE Code of Conduct, covering

(*a*) Integrity
(*b*) Competence
(*c*) Interests of the public
(*e*) Continuing professional development
(*f*) Legal compliance
(*d*) Environment and sustainability.

I have rearranged one of them so that the initial letters form an acronym – ICICLE – thus making them easier to memorise.

The Institution goes into some detail in the preamble about

- the purpose of the Code
- what ICE sees as the overriding ethical duty of its Members
- how it is able to support anyone who is troubled by an ethical problem.

Following the six rules are Guidance Notes, explaining what the rules entail. Interestingly, Rules 2 and 5 have explanations expanding on the rule itself while, in contrast, there are detailed explanations of how Rules 1 and 3 could be broken. Rules 4 and 6 are considered adequate as they stand, probably because there is sufficient explanation elsewhere, such as MGN 24 on Sustainability. It is well worth reading and understanding both the preamble and this guidance.

In the unlikely event that you do call upon the Institution for advice on ethics, do remember that it is a professional body and not a trade organisation. They will not (cannot) take your side, but will offer a balanced view of the situation, leaving you to decide whether to pursue your stance or not. This can prove frustrating if you are looking for committed support. Only if you have an 'open and shut' case will you get that support, which you are probably desperately seeking.

ICE *Advice on Ethical Conduct*

Part of the advice to members is a document *Advice on Ethical Conduct*. This is rather like the UK's Highway Code: if you are accused of breaking a Rule (the law), then any breach of the Advice (the Highway Code) can be used in evidence against you, but the advice is not, of itself, obligatory (see Chapter 9). This advisory document is well worth studying: it not only sets out clearly what behaviour is expected in a variety of circumstances, including those that cause most complaints against Members (which explains the inclusion of party walls), but appends to each section a useful and relevant list of further reading. As I recommended in Chapter 10, it is important that you know the duties and responsibilities of yourself and others *before* any problems arise.

RAEng *Statement of Ethical Principles*

The Institution has also endorsed the *Statement of Ethical Principles* first drawn up by the Royal Academy of Engineering in 2005. It contains four principles, summarised as:

- Accuracy and Rigour
- Honesty and Integrity
- Respect for Life, Law and the Public Good
- Responsible Leadership, Listening and Informing.

I hope that, by now, you have realised that, in a sense, this whole book is about the initial development of a totally professional attitude. It covers the basics of professional commitment – the wide range of knowledge and understanding needed to make holistic judgements for the benefit of mankind and the environment.

Rigid ethical adherence to a code of morality is not easy! In my own 50-year career, I have been summarily dismissed (reinstated two hours later after the intervention of a solicitor), have resigned (resignation not accepted, but transferred to another role) and have been called to disciplinary meetings. It does sometimes seem that the easier option is to give in but, in the longer term, I was always vindicated and did, I think, gain some grudging respect. What I would counsel from this experience is that you should never do something on the spur of the moment through instinct or gut feeling, but prepare a detailed factual case first. You may not need it, but at least you then know that you have a well-argued and specific case which your accusers will have difficulty in repudiating.

Continuing Professional Development

I cannot see any way in which a competent civil engineer can make career progress and keep up to date with the ever-changing plethora of legislation and technical standards, let alone monitor the changing public criteria for infrastructure development, without continuing their professional development. I think the difficulties many have had with formal CPD is in the recording, rather than in the doing. It is almost second nature and perhaps many of us do not realise we are doing it virtually all the time.

There are those, too, who believe that because they are professional civil engineers then, *ipso facto*, they must be keeping up to date and competent. This may well be true, but the reality is that, with the present UK 'blame culture' and adversarial legal system, any one of us could be called to account for our decisions, and the prosecution will do almost anything to discredit us. A comprehensive written record of a serious attempt to keep up to date with technology, the law and public attitudes must surely be essential to our defence in this scenario.

The Institution also requires us to anticipate what we may need to know in the future, by drawing up a Development Action Plan (DAP). For those in the early stages of their career, this requires them to decide on their ambitions and career plans – what they need to learn for the next stage of their advancement, as well as keeping up to date with developments in their current role.

The Institution's random checks on a percentage of Members' CPD is no more than an attempt to ensure that every Member's 'back is covered' in the event that their competence is questioned. Having had the experience of several court appearances as a witness, I shall always be grateful to those bosses who made me (albeit reluctantly early in my career) keep records of my professional development. Since then, my written CPD record has consistently and easily exceeded the Institution's stated minimum – and I am sure that, even now, I do not remember to record everything I read or learn.

Purpose and history of the ICE

The Institution of Civil Engineers has a long history over nearly 200 years. Nearly every potential Member can recite its beginnings in Kendal's Coffee House, but I fear that few know much about its recent history, nor indeed why it exists. The Head of Knowledge Transfer at the Institution, currently Mike Chrimes, has put much effort into rectifying this shortfall, and his talks are well worth attending or downloading from the Institution's regional websites.

Only relatively recently (since the Second World War) has the Institution gone through really major changes, starting during the period when the Secretary was Garth Watson and continuing at an increasing pace ever since. Major changes to its governance and the manner in which it conducts its business have affected, and continue to affect, all members, so everyone should have some understanding of what these changes are.

Nowadays, the Institution routinely surveys the views of its membership to ensure that it is complying with what the Members want from their professional body. From these surveys, they have concluded that the Institution needs to

- be more outward looking
- emphasise ICE's impact on society – around the world
- develop strong positions on major policy issues facing the profession and society (through the Learned Society and Knowledge Strategy)
- understand key external stakeholders and develop their relationships (e.g. working with other institutions/bodies)
- understand employers' needs and expectations
- improve business efficiency – more integration across the ICE Group with coherent data/web/IT capability
- ensure a coordinated ICE embracing Regions, Centre and International
- invest in their staff and volunteers.

The Institution wishes to be a leader in shaping the engineering profession, particularly in construction and related activities. It has decided to achieve this by

- being broad and inclusive of all those engaged in civil engineering
- maintaining high professional standards for qualification and membership
- providing an international source of knowledge and skills for tomorrow's engineers
- using its knowledge and skills to deliver products and services that are attractive, relevant and valued by Members and employers globally
- promoting the economic role of civil engineering and raising the profile of civil engineering
- influencing governments, industry and other stakeholders to value civil engineering
- partnering with relevant bodies.

Everyone aspiring to become a Member of ICE should have an outline understanding of 'current developments affecting the construction industry' and the role that the Institution has in resolving the huge 'issues' arising from them.

Promotion of the construction industry

Civil engineering works are mainly public-sector promoted, so politics largely governs the planning and expenditure on infrastructure provision and maintenance.

In my considered view, civil engineers should avoid joining the political debate, even if employed by a public authority. It is the civil engineer's duty to provide unbiased economic, social, environmental and technical information to allow political decisions to be taken in full cognisance of all the facts. Defining the consequences of each of a likely range of available decisions (the political options) is an intrinsic part of that advice. In addition, civil engineers are well equipped to provide advice on the hierarchy of priorities, by using their asset management skills.

Having said that, I have colleagues who became elected politicians. They seem to have few problems in reconciling their political work with their professional standards. After many hours debating with them, I do think, however, that there is a fundamental difference between us. They are *employed* as politicians but bring to that job the judgement, knowledge and professional responsibility of a civil engineer. They are neither employed to provide the professional judgements of a civil engineer nor to deliver the best possible outcome from their political decisions. There is therefore no clash between their political standards and the Code of Conduct.

Once a political decision is taken, it is the civil engineer's duty to fulfil it in the best possible manner (Royal Charter 'use of scarce resources, care for the environment and in the interests of public health and safety'), regardless of any personal views.

The Institution as a corporate body is possibly in the best position to give advice to the government on national and international policy on construction. It is largely perceived to be an unbiased organisation, of great value to politicians of all persuasions because it is able to draw on worldwide expertise, including its Local Associations and Specialist Engineering Groups. But there is always a danger of being perceived, whether justified or not, as a pressure group or lobby. This can be a particular problem when an individual becomes closely identified in the public perception with a specific topic. Any such perception is detrimental to the continuing political worth of the Institution's unbiased professional advice and must be avoided at all costs. It is a difficult balance, but the whole of our professional development trains us to make careful judgements and we should be able to get it right.

Each year, the Institution publishes a 'State of the Nation' Report, gathering material from the UK Regions into a concise document of use to all those engaged in trying to decide on strategic directions and funding priorities. It is surely reasonable for the Institution (through its Reviewers) to expect potential Members to know what the most recent report highlighted, both on national and regional levels. Have you or your employer ever contributed and, if so, how?

At a more local level, there are many opportunities for the individual civil engineer to influence and contribute to debate about infrastructure, by becoming involved with, for example

- Chambers of Commerce or Trade
- Round Table
- Civic Trust
- clubs and societies.

It is possible for the individual civil engineer to influence society's views by assisting with

- industrial liaison on school/college/university projects
- schools liaison
- Neighbourhood Engineer
- ICE Ambassador scheme
- CITB presentations initiative
- governance of a school or college

and by giving talks to clubs and associations on topics of local interest, such as

- Women's Institute, National Women's Register
- Working Men's Clubs, etc.

At these, it is important to remember the responsibility to present the unbiased facts, keeping personal opinions discretely in the background.

Another useful way of raising the profile of the profession is through charity work, such as raising sponsorship for, or participating in, causes with an engineering content, such as

- Water Aid
- Overseas Aid
- Register of Engineers for Disaster Relief (REDR).

So, every individual civil engineer does have opportunities beyond the workplace to demonstrate 'the special skills and professional approach' of our profession. As you have gone through this book, I hope that you have begun to realise what these are, and how you can develop them as efficiently as possible within your own unique working environment.

Initial Professional Development for Civil Engineers
ISBN 978-0-7277-4147-9

doi: 10.1680/ipdce.41479.113

References

Baker, J.A. *et al.* (2007) 'Panel statement', *The Report of the BP US Refineries Independent Safety Review Panel.*

Business Round Table New York (1983) *More Construction for the Money*, Summary Report of the Construction Industry Cost Effectiveness Project.

Chitty, J. (2010) *Chitty on Contracts*, London: Sweet and Maxwell.

Defra (2005) *Securing the Future: UK Government Sustainable Development Strategy*, UKCDS.

Department of Environment (1988) Circular 15/88. www.planningsanity.co.uk/forums/legal/circulars/15-88.htm (accessed 25 November 2010).

Egan, Sir John (1998) *Rethinking Construction*, London: HMSO.

Gilbert-Smith, D.S. (2003) *Winning the Hearts and Minds: A Book on Leadership*, Brighton: Pen Press Publishers Ltd.

Kipling, R. (1902) 'The Elephant's Child', *Just So Stories.*

Latham, M. (1994) *Constructing the Team*, London: HMSO.

Macdonald Steels, H. (1994) *Effective Training for Civil Engineers*, London: Thomas Telford.

New Civil Engineer (2007) Definition of civil engineering. www.nce.co.uk/new-definition-of-civil-engineering/212786.article (accessed 24 November 2010).

Institution of Civil Engineers (1999) *ICE Conditions of Contract – 7th Edition*, London: Thomas Telford.

Institution of Civil Engineers (1995) *New Engineering Contract*, London: Thomas Telford.

Institution of Civil Engineers (1996) *Sustainability and Acceptability in Infrastructure Development*, London: Thomas Telford.

Priestley, J. (1794) *Heads of Lectures on a Course of Experimental Philosophy*, www.josephpriestleyhouse.org/index.php?page = priestley-and-education (accessed 24 November 2010).

Prospects 'Graduate Careers', www.prospects.ac.uk/options_civil_engineering_your_ skills.htm (accessed 24 November 2010).

Short, C. (2004) *An Honourable Deception?: Labour, Iraq, and the Misuse of Power*, London: The Free Press.

Walker, T. (2005) 'Difficult decisions – HSE's approach', Speech, HSC/E Seminar on Risk and Compensation – Striking a Balance, 22 March 2005. www.hse.gov.uk/risk/timothywalker.pdf (accessed 25 November 2010).

Wolstenholme, A. (2009) *Never Waste a Good Crisis: A Review of Progress since Rethinking Construction and Thoughts for Our Future*, Constructing Excellence.

World Commission on Environment and Development/Brundtland Commission (1987) *Our Common Future*, Oxford: Oxford University Press.

World Health Organisation (1993) 'Draft definition of "health"', Sofia, Bulgaria, available at www.health.gov/environment/definitionsofEnvironmentalHealth (accessed 25 November 2010).

Legislation

Construction (Design and Management) Regulations 2007. www.hse.gov.uk/construction/cdm.htm

Health and Safety at Work etc. Act 1974. www.hse.gov.uk/legislation/hswa.htm

Management of Health and Safety at Work Regulations 1999. www.legislation.gov.uk/uksi/1999/3242/contents/made

Management of Health and Safety at Work (Amendment) Regulations 2006. www.legislation.gov.uk/uksi/2006/438/contents/made

Useful websites

www.businessballs.com

www.citation.co.uk/riskassessments.asp

Engineering Council 'Engineers and the Environment'. www.engc.co.uk/about-us/sustainability

ICE Code of Conduct. www.ice.org.uk/mgn43

ICE Engineering Principles. www.ice.org.uk/mgn47

Manual Handling Operations Regulations. www.legislation.gov.uk.uksi/1992/2793/contents/made

Personal Protective Equipment at Work Regulations. www.legislation.gov.uk.uksi/1992/2966/contents/made

RAEng Statement of Ethical Principles. www.raeng.org.uk/societygov/engineeringethics/principles.htm

'Rethinking Construction' Executive Summary. www.berr.gov.uk/files/file14364.pdf

Royal Charter. www.ice.org.uk/About-ICE/Who-we-are/Royal-Charter-By-laws-Regulations-and-Rules

Sustainability and Acceptability in Infrastructure Development (SAID). www.icevirtuallibrary.com/content/book/100952

United Nations Framework Convention on Climate Change (Kyoto Protocol). http://unfccc.int/resource/docs/convkp/kpeng.html

Initial Performance Development for Civil Engineers
ISBN 978-0-7277-4147-9

doi: 10.1680/ipdce.41479.115

Index